Loosening the Bonds

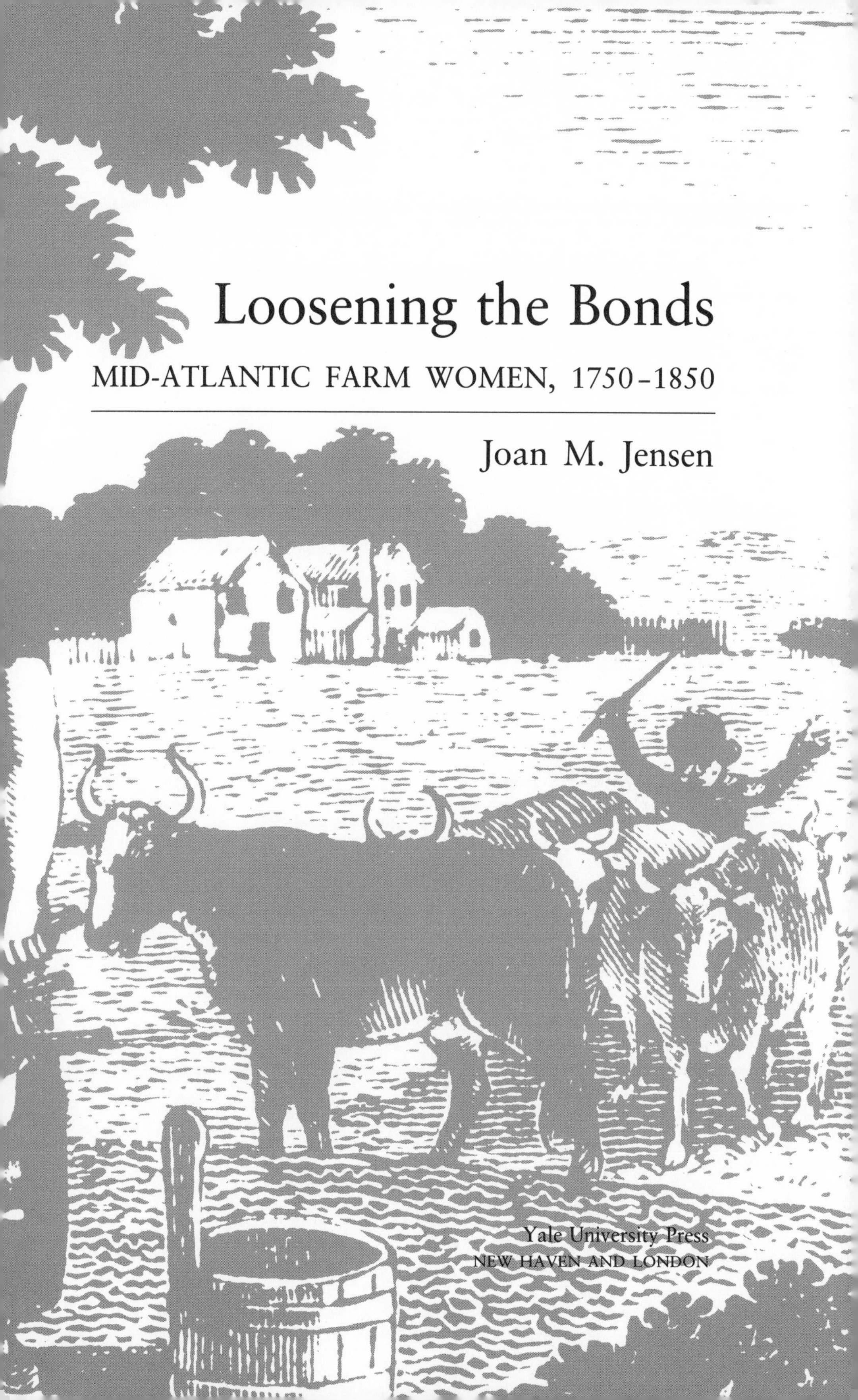

Loosening the Bonds

MID-ATLANTIC FARM WOMEN, 1750–1850

Joan M. Jensen

Yale University Press
NEW HAVEN AND LONDON

Designed by Susan P. Fillion
and set in Sabon type by Rainsford Type.
Printed in the United States of America by Murray Printing Company, Westford, Massachusetts.

Library of Congress Cataloging-in-Publication Data

Jensen, Joan M.
Loosening the bonds.

Includes index.
1. Rural women—Middle Atlantic States—History.
2. Women in agriculture—Middle Atlantic States—History. 3. Women in public life—Middle Atlantic States—History. I. Title
HQ1438.A12J46 1986 305.4'2'0974 85–22471
ISBN 0–300–03366–4 (cloth)
0–300–04265–5 (pbk.)

The paper in this book meets the guidelines for permanence and durability of the Committee on Production Guidelines for Book Longevity of the Council on Library Resources.

10 9 8 7 6 5 4 3 2

For ROSALIA GUSTINIS, 1910–1983,
refugee Lithuanian farm woman who helped me
with this book in more ways than she knew

Contents

Illustrations

ACKNOWLEDGMENTS

So many historians, archivists, curators, folklorists, and friends helped me along the way that I can only remember, not list, the ways in which they made this book possible. A few stand in need of special thanks, however. I would particularly like to thank Jack Michel for making it possible for me to spend a summer at his eighteenth-century Brandywine Valley farm. There the morning snorts of the horses, the rain-soaked siding of the bank barn, and the quiet garret enabled me to sense what life might have been like for the women of the valley. Saez Fussel Macauley, who welcomed me to her home to use and discuss the Esther Lewis papers, encouraged me greatly. Later she guided me to the two Lewis homes in Vincent Township, a visit that made this nineteenth-century farm woman a real person for me. The friendship of these people helped make the sometimes lonely and often tiring hours of research an exciting adventure. I also remember Sunday picnics with Mary and Philip Johnson at the Brandywine River State Park, where we talked about Quaker culture for hours, and other picnics with Nan Adshead at the old New Warke meetinghouse and burial ground. Lucy Simler helped me understand the intricacies of Chester County documents. Gary Nash gave me support at all stages of the project, from the initial proposal, when others were skeptical, to a critical reading of the final manuscript.

During the last few months, when finishing the manuscript became a grueling chore, two women made my surivial possible. Holly Reynolds, the Interlibrary Loan librarian, found all the last obscure references I still needed and Monica Torres helped me with the final revision. I truly thank them, for finishing is always more difficult than starting. And last, I would like to thank John Gustinis, who spent his first ten years on a Lithuanian farm, from all accounts much like those of the Brandywine Valley. The oral histories of his parents, along with his memories, provided many clues about what to look for in the records of the past. His first enthusiastic reading of the manuscript gave me hope that I might have grasped something of what it was like to live on these early farms.

Parts of this research were generously funded by grants from the Regional Economic History Research Center at the Eleutherian Mills–Hagley Library, by the Philadelphia Center for Early American Studies, and by a sabbatical from New Mexico State University. Chapter 10 appeared in slightly different form as "Not Only Ours But Others: Quaker Teaching Daughters of the Mid-Atlantic, 1790–1850," *History of Education Quarterly*, June 1984.

INTRODUCTION

This is the first book to focus on the lives of women of the rural majority in the critical century that spanned the late colonial and early national eras. Drawing on a wide range of sources and methods, I have attempted to show the complex lives of rural women in the Philadelphia hinterland during it's rise to prominence as one of the wealthiest agricultural regions in the country. The tools women used, such as churns, the letters they wrote, and the brief records left by others provide a rich textural background. There was hard work in the lives of these women—in field, in dairy, and in farmhouse—but there was also an adventurous exploration of new roles as ministers, teachers, and reformers. The area of southeastern Pennsylvania and northern Delaware on which this study centers was a Quaker culture area. Nonetheless, women of diverse Euro-American and Afro-American families worked, suffered, and thrived on the fertile soils of this area. Each in her own way worked to loosen the bonds of control by family and community over her life and to create a new fabric of society.

Behind the bustling commercial port of Philadelphia, largest city in the English colonies, lay a rich rural hinterland. Cut by the Schuylkill and Brandywine rivers, the rolling hills spread for miles, dotted with farmhouses and fields marked by wandering worm fences. In 1750, many acres were still forested and the hinterland was a frontier. By 1850, the farms were settled, the forests had been cut back, and the farm families had learned to survive not frontier harshness but the harshness of competition in a world of urbanizing and industrializing capital. They had made the transition to a new commercial farm system by developing new skills and new market products.

This study began as an attempt to tell the story of what happened to the women who remained on the farm during this momentous transition in American history. Many histories discuss the urbanization and industrialization process of the new nation; some even discuss the role of women in that process. None focuses on rural women and their place in the transformation of the American agricultural system. These women did not become mill girls, nor did they become the models of "true womanhood" who reigned over homes left vacant by men drawn into the new industrial economy. They worked with men and children in a farm family household, essentially unchanged in its basic form of property ownership and work relations. But neither their work nor their lives remained unchanged. This is the story of how that work and those lives changed and became essential parts of the agricultural transition of the early nation. It places the rural majority in its proper place, at the center of the history of American women.

The first sphere of rural women was the household. Through wills and inventories, poorhouse and church records, rural women emerge as active participants in shaping the late eighteenth-century culture of the Mid-Atlantic colonies. These records allow complex questions to be answered. What were the community, family, labor, and welfare structures of women? Were rural women active participants in the passing on of property between generations? Did they work in the fields? If they did not sow, did they still reap? Who cared for the working poor who had no kin in an era of kin-based welfare? What happened when a farm servant became pregnant and there was no father to share in caring for her expected child? What were her options? How did the community care for her? Who were her fellow inmates at the poorhouse? Why were they there?

Rural women's second sphere was the domestic production they performed for the market. In some ways, this commercial sphere was the most crucial, for through it women directly shared in economic development. Here, the focus is primarily on what happened to rural women's work during the industrialization process. The women of the Philadelphia hinterland moved into and developed a new cottage industry that provided an economic transition for farm families. These families were responding to a volatile commercial market, itself affected by the productivity of new lands to the west. Wheat, pigs, sheep—traditional farm crops of the male—were more easily and cheaply raised in the West. The butter that farm women had traditionally made now moved to a central place on these farms. The churn came to symbolize not the domestic arts of housewifery but the commercial arts of women alert to the demands of the market. In this rural transition, what happened to ideas about women's role in the rural economy? Almanacs and treatises on agricultural development give us new ways of looking at that question. "You may be sure, she eats no idle bread," wrote Pennsylvania farm woman Esther Lewis in the late 1830s, referring to the young servant who worked with her to produce butter for the market. Such work kept the farm household a functional, fundamental, and essential part of the new economy. In this period of newness and nation building, then, women participated not only in the household but also in the marketplace.

Rural women's third sphere was public. Women emerged into the public sphere through their activities in religion, education, and reform. Whether Quaker ministers learning how to be "Public Friends," young teachers trying out a new independence working in places as distant as Alexandria, Virginia, or reformers appealing to sisterhood as a new bond to replace hierarchial relationships, these women exhibited extraordinary energy and imagination as they moved self-consciously out of the private and into the public sphere. They did not emerge free and equal, but they emerged freer and more equal into a society that would try their skills and challenge their ambitions. Because these women worked in three spheres, their culture became rich, dense, and intricate. It bound them to husband, children, neighbor, market, and to other women. But the changing culture also loosened those bonds.

In her analysis of New England women, Nancy Cott used the term *bonds of womanhood* to describe a dual condition, both the subordination women experienced and the sisterhood they forged to escape from that subordination. I use the term *bonds* in primarily the first sense, to describe the ways women were held in a subordinate place in family and community.

This study suggests the diversity and change in these human and material relationships. It offers a new view based not on one but many methods, for no culture as complex as that of these farm women could possibly be re-created without exploring many interdisciplinary paths. These early rural women have to speak through unorthodox documents—the records of pounds of butter processed, stained wooden churns, crabbed notes in ledgers, Xs signed to wills, and precisely written signatures on petitions. Their voices speak clearly nonetheless.

Chester County in Pennsylvania and New Castle County in Delaware are the microcosm through which this study explores the lives of these women. Joined geographically in the Brandywine Valley, cut politically by the Mason-Dixon Line, a Quaker culture area where abolition, temperance, and women's rights activity flourished—this Philadelphia hinterland is an ideal area in which to begin the process of recovering the history of rural women.

PART I

THE HOUSEHOLD

CHAPTER 1

Culture on the Brandywine

Today, as in 1750, the farmlands of the Brandywine Valley stretch mile after mile through the Philadelphia hinterland. From its source in the Welsh Hills, sixty miles and three thousand feet above the tidewater, the Brandywine River flows southward to join the Delaware River at Wilmington. It cuts through what farmers once called the Great Valley—a trough between the Schuylkill and the Susquehanna rivers. The river forms this smaller valley, joining Chester County in Pennsylvania and New Castle County in Delaware, into one geographical unit. Much of the area is still forested—actually reforested—giving it an appearance similar to what it had when eighteenth-century women of the Brandywine Valley looked across the low rolling hills to the muddy river waters.

Underneath the trees and the crops, now mostly corn and hay, the soil is still crystalline shale, sandstone, and limestone. The winters are usually mild, although they can be harsh. There is often snow from December to early March, as much as eight to thirteen inches. In March come the harbingers of spring: the chattering return of the bluebirds and blackbirds, then the cooing doves, the emergence of frogs, ground squirrels, and snakes. April is a sudden bursting of spring with blossoming trees, their scent followed by the pungent smell of wild garlic, now as then rampant in the fields. The growing season of 180 days gives ample time for corn to mature. The forty-six inches of rainfall a year makes the land lush, especially in spring when creeks run full and later in July when rainfall is heaviest. Then follow stifling and humid days, often with no winds, a time when people and animals suffer but the crops flourish. Some years are especially dry. In 1755–56 the streams dried up. Later in the eighteenth century and again in the 1830s there were droughts. Usually, however, the heat ends within a week as rain drenches the valley again. August brings thundering storms. And sometimes in September there are fogs such as Elizabeth Drinker described in 1796 when she wrote in her diary: "They will rise morning and evening in the meadows about a yard high and look just like a field of buckwheat in blossom."[1]

The Brandywine Valley settlements were no longer sparse by 1750. Mills were established along the lower river a few miles north of Wilmington, in a small settlement to which farm families brought their surplus wheat to exchange for imports. For over a decade, Brandywine millers had been exporting this rural surplus. At annual fairs in April and October country folk also bartered and sold. Most rural women and men traveled by horse as few could afford carts or wagons. Although land was widely distributed among the Euro-American yeoman families, distinct classes were already evident in the substantial stone houses the wealthy were building and in their displays of wealth. Three generations before, Anglo-American Quaker settlers had replaced many earlier Swedish and Dutch settlers, making theirs the dominant culture of the Brandywine Valley.[2]

The first Europeans to settle among the Lenai-Lenape Indians were Swedish. They bought small parcels of land and for some time traded and peacefully coexisted. Indian women at that time still left their villages on the banks of the Brandywine to cultivate fields of corn and gather food. Swedes soon saw their collecting as stealing, however, and conflict increased as women harvested in orchards set out by Swedes. The Swedes, as the traveler Peter Kalm put it in the 1750s, "gave them a severe drubbing, took the fruit from them, and often their clothes, too."[3]

Native Americans also developed a brisk trade in venison and corn with the first British immigrants. Quaker women in many parts of Pennsylvania had good memories of these early relations with the Indians. One Bucks County family asked neighboring Indians to check in on the children while the adults went to meeting. Mary Smith, a West Jersey Quaker settler, learned their languages and became an interpreter. But British diseases proved more fatal than Swedish drubbing. As Mary Smith wrote in the early eighteenth century, "God's providence made room for us in a wonderful manner, in taking away the Indians. There came a distemper among them so mortal that they could not bury all the dead. Others went away, leaving their town."[4]

With the help of providence, the banks of the Brandywine were soon cleared as well. All but one of the native inhabitants, Hannah Freeman, had died or moved on by 1750. Hannah collected herbs and seeds, sold herbal remedies, brooms and baskets of oak, and ash splints from farm to farm. By the 1780s, she was known as "Indian Hannah" and she appears in the record books of a Birmingham Township farmer as buying corn, wheat, and cider. Old-timers remembered Hannah as saying her people caught great quantities of fish on the Brandywine with nets made of grape vines and that her people had been wronged.[5]

By 1763, almost all the Delaware Indians had died or begun their westward trek. As Anthony F. C. Wallace concludes, they "had been elbowed out of the last of their ancestral homeland" and those not reduced by massacre or disease "remained as tenants on the land that had once been theirs." Some settled in western parts of Pennsylvania; other crossed into what would become Ohio. The culture was so disrupted and has been so little studied that scholars still disagree as to whether it was matrilineal and matrilocal or patrilineal and patrilocal before European contact. Less than one hundred years after the beginning of European settlement, it was a culture in flight.[6]

Those Indians who left were able to maintain a cultural identity. The Swedes who stayed did not. By the 1750s, the Swedes were undergoing a demographic transition that resulted in their virtual disappearance as a cultural entity. Malaria and other diseases carried off many of the older Swedes as they had the Native Americans. Instead of moving west, however, the remnants stayed scattered in the Philadelphia hinterland, their children gradually being assimilated into the English culture. The families continued to be farmers—a 1753 census of the Swedish church still called almost 90 percent of them "peasants." The rest ventured singly into commercial or artisan occupations. The Swedish, who defined themselves primarily by language in the seventeenth century, absorbed many of the early Dutch and remained a living culture into the eighteenth century. After 1750, however, few spoke Swedish or identified with the Lutheran church. Men, particularly, married English wives who spoke no Swedish and brought up their children as English. Lower fertility, higher adult mortality, and declining ethnic-linguistic identity gave the Swedes only slightly better survival chances than Native Americans. "The English are evidently swallowing up the people," wrote one disturbed Swedish minister visiting nearby West Jersey communities in 1745. The same was happening to those Swedes settled on the other side of the Delaware River in the Brandywine Valley.[7]

What Native Americans and Swedes lost, English Quakers gained. The swelling Quaker migration of the late seventeenth century into the Philadelphia hinterland and the Brandywine Valley continued into the early eighteenth century. But belonging to the Quaker community did little to preserve the ethnic culture of a group, as the experience of the Welsh Quakers demonstrates. Immigrants from Wales scattered through the Swedish and English population, mainly in northeastern Chester County. The Welsh Quakers settled as a group in a forty-thousand-acre part of the Philadelphia hinterland known as the Welsh Tract. Early founders of these communities expected to establish Welsh Quaker culture there, but they soon realized the difficulty in maintaining ethnically plural communities within the Quaker structure. The yearly meeting was controlled by coalitions of wealthy English rural and urban Quakers who discouraged ethnic celebrations, insisted on the use of English in meeting records and publications, and defined religious unity as acceptance of their unique form of English culture. Therefore, in practice, the fate of the Welsh was little different from the Swedes. They too were "swallowed up" by the English and their most visible customs disappeared by the eighteenth century.[8]

It is possible, of course, that once historians have developed more sophisticated methods of examining regional variations within the Quaker culture, they will find more than the Welsh place names of Radnor, Gwynedd, Tredyffrin, and Uwchlan and a cluster of family names derived from the Welsh, such as Lewis, Thomas, David, and Williams. Perhaps careful analysis of architectural styles and other aspects of material culture will reveal ethnic patterns that coexisted within the dominant English culture. At the most visible level, however, Welsh culture had disappeared by 1750. Only one of the hundred wills I examined from the 1750s for this study revealed any trace of the culture. Mary Davis willed her Welsh Bibles and books to her son John in 1753. That Bible was a last reminder of language and culture given up by Welsh women in return for membership in the larger Quaker community.[9]

The Irish Quakers, who scattered along the banks of the Brandywine River, like the Welsh left little visible record of a separate culture. English Quakers migrated to Ireland in the seventeenth century where they became prominent in the textile trade, in retailing, and in shipping. They also converted native Irish and numbered perhaps ten thousand by the eighteenth century. An estimated fifteen hundred to two thousand of these Irish Quakers migrated to Pennsylvania in the late seventeenth and early eighteenth centuries. Most of them spoke English, observed English social customs, and, as descendant Gerelyn Hollingsworth has noted, considered themselves English, little changed after several generations spent in Ireland. What did emerge as a separate identity was the political involvement of the men (many of whom had previously held public offices in Ireland), their intense interest in financial matters, and opposition to the proprietary government.[10]

Those Irish Quakers who left Philadelphia concentrated in the Brandywine Valley, in both New Castle and Chester counties. Of all the hinterland townships (the smallest administrative unit of Pennsylvania counties) in southeastern Pennsylvania, Kennett and New Garden received the largest influx of Irish Quakers. Some settled on the Delaware side of the border in the Christiana and Brandywine hundreds (the smallest administrative unit of Delaware counties) as well. Irish Quaker women came with their husbands and children or, if they were single, married other settlers. Women born along the Brandywine took Irishmen as their husbands. The Irish, according to Albert Myers, preserved a strong separate identity, but that identity had almost disappeared by the mid-eighteenth century.[11]

Women played an important role in maintaining the family alliances that gave a strong base to the Irish minority during the early eighteenth century. A good example is women's role in the Hollingsworth and Harlan families. Ann Calvert Hollingsworth came from Ireland with her husband Valentine in 1682. Their four children either came with them or followed within a few years. Ann bore three more children near the banks of the Brandywine before she died fifteen years later. Elizabeth Dick Harlan came from Ireland in 1687 with her husband George and four children. She bore five more in the next nine years. In 1701, Elizabeth's oldest daughter, then twenty, married Ann Hollingsworth's oldest son, then twenty-eight. By 1751, however, Ann's son was dead as were all of Ann's other children except Valentine. And all the children, including Valentine, married English women.[12]

The marriage of Valentine Hollingsworth II to Elizabeth Heald was, in many ways, representative. Elizabeth Heald's parents came from Cheshire, England. Two of her sisters, Sara and Mary, married two of Hannah Harlan's brothers, Aaron and Joshua Harlan. Intermarriage was common because many of the Irish Quakers had already intermarried with the English or had come from England themselves only a generation or two earlier. Although the Healds, Hollingsworths, and Harlans, continued to intermarry in an increasingly complex genealogical pattern, they retained little overt Irish ethnic identity. A few continued to live around Centreville and north of Wilmington at New Warke, but most families scattered, carrying on an exogamous dispersal pattern already established in Ireland. Many children also married English Anglicans. Five of the seven children of Sarah Heald and Aaron Harlan were married in Episcopal ceremonies, including twenty-three-year-old Elizabeth Harlan who married Valentine Hollingsworth III in 1743.[13]

Thus, as the Irish increased and multiplied along the banks of the Brandywine, they often lost both their Quakerness and their Irish identity. Although a few of the older Hollingsworths and Harlans—male and female—died illiterate in the early eighteenth century, they left carefully dictated wills. Sarah Harlan, for example, left large quantities of cloth and four spinning wheels. The Irish Quakers left an immense legacy of material culture—in pots, featherbeds, clocks, looking glasses, pewter, chairs, tables, and linen cloth. They also may have bequeathed a nonmaterial culture. Myers described that legacy as a tendency in the Quaker meetings where they settled to be more liberal in belief and less stringent in the administration of some of the rules of discipline. But, overall, Irish Quakers left less of a separate culture in Chester County than even the Welsh.[14]

Nor did non-Quaker Irish leave a visibly separate culture, although they too were scattered throughout the Brandywine Valley and Chester County by 1750. James Lemon estimates that perhaps 23 percent of Chester County was Scotch-Irish by 1759. There was one stronghold of Scots-Irish Presbyterianism near New Castle, Delaware, and some moved north across the Pennsylvania line in the late seventeenth century. After 1700, hundreds of Scots-Irish moved into Chester County, forming four townships on the east bank of the Susquehanna River. They also established farms on the Brandywine River and on the Elk and White Clay creeks in New Castle County. The Scots-Irish probably were the first to cultivate the Irish potato, which became a staple in the Mid-Atlantic region and spread throughout the colonies. They also left place names and family names in the area. After the 1740s, additional Scots-Irish Presbyterians, who had left northern Ireland because of difficult economic conditions, moved into the hinterland looking for farms, some bringing with them native Irish Catholic servants. Presbyterians formed eight churches in western and southwestern Chester County before 1750 and Catholics formed a church in West Chester in 1793, a reflection of the importance of both Scots-Irish and native Irish immigration into the area. But they formed only small islands among the Quaker majority.[15]

Germans, on the other hand, were able to maintain a separate identity. But those Germans who entered the valley moved north to settle in dispersed communities, bought a few unsettled parcels of land, and established strong cultural ties with German communities to the north. Others who arrived too late to purchase land cheaply pushed on to the western frontier of Chester County to settle in what became Lancaster County, leaving a deep cultural mark there rather than in the Brandywine Valley.[16]

The French never formed a distinctive community. A number of French became Quakers—Anthony Benezet was the best-known. Others married Anglo-American women and adopted their culture. Most French, in any event, settled in Philadelphia. A few French families established farms, but in Montgomery rather than in Chester County. One French bachelor who farmed on the Schuylkill River hired a German couple to help him with his 250-acre farm. Refugees from the West Indies and from revolutionary France tended to choose urban settlement or, like the du Pont de Nemours, brought industry to the banks of the Brandywine River and farmed only as a sideline.[17]

The only culture strong enough to retain a unique ethnic identity in the predominantly English Quaker culture area was that of the African-Americans. As slaves, blacks were under great pressure to assimilate, but as blacks became free after the 1770s, they began to form their own communities. The color line established by the Quakers, combined with a strong non-European cultural heritage, impeded blacks' assimilation. As the free black population increased in the late eighteenth century, Quakers hesitated to accept even racially mixed people who had adopted the Quaker culture. For a time, blacks seemed to find a home with the Methodists, whose religious style more closely matched their own. But by the late eighteenth century, blacks were seeking to establish separate congregations, and when Methodists began a policy of segregation, a large proportion of blacks responded by seceding. Along Brandywine Creek, at least one community had formed by 1779 that was composed of blacks who were employed by wealthy neighbors and, according to a Quaker visitation committee of that year, lived "well as to the things of this life." Other blacks fared less well. In any event, as black communities grew, they grew separately.[18]

With the settling of the small black population, the ethnic composition of the valley was fixed until the second wave of immigration from southern Ireland in the 1830s. English people, who settled throughout the valley, predominated everywhere except in those few areas settled by the Germans. Irish Quakers remained concentrated in the lower valley, Welsh Quakers to the north and east, with very small numbers of Dutch, Swedish, and blacks scattered through the rest of the valley. The valley, however, came to be dominated not by an English but by a Quaker culture. Although religious diversity was a hallmark of the Mid-Atlantic region, the Friends' meetings far outnumbered all other denominations (see Appendix table 1).

Like the Yankees in New England or the Germans in western Pennsylvania, Quaker farm families developed a distinct culture in southeastern Pennsylvania. Sidney George Fisher, a Philadelphia lawyer who owned a farm in nearby Maryland, remarked in his diary of November 8, 1839, on the unique place Quaker culture occupied in the Mid-Atlantic. The farmers of Maryland, he said, were different from the Yankees, the Germans, "or the plain, homely, steady population of our Quaker countries, who with all their faults, are moral, industrious and quiet, and cultivate with great skill their fertile valleys."[19]

The Brandywine Valley Quaker culture first centered in the New Warke Monthly Meeting, north of Wilmington. Today, only the old meetinghouse with its wall-enclosed burial ground remains as a reminder of the once burgeoning religious community. Valentine Hollingsworth set aside land for the meetinghouse and burial ground in 1690, and sixty years later, in June 1750, the meeting had reached its largest membership. In August of that year, the crowded meeting divided, and the Wilmington Monthly Meeting provided a new focal point for the growing number of Quaker families in and around that village. By 1760 New Warke

Monthly Meeting itself had ceased to exist and became part of Kennett Monthly Meeting. Thus 1750 provides an ideal point at which to examine the New Warke community.[20]

This religious community numbered over 450 individuals in 1750. Unfortunately, the complete community cannot be reconstructed because records do not exist for all the older members or the less active members. However, by reassembling extant documents, which are amazingly abundant for this period, the researcher can re-create a partial census for the New Warke community. Slightly more males than females made up the membership, but the gender imbalance was small. Children under ten made up 27 percent of the community, with sixty of each sex. Another 11 percent were youths between the ages of ten and seventeen, 8 percent were between eighteen and twenty-five, and 15 percent were over twenty-five. The age of 39 percent of the adults could not be determined. But, if one assumes their age to be eighteen or over, then approximately two-thirds of the church community would fall within that age group. At least twelve women were pregnant in 1750 at the time of our reconstructed census. Almost half of the women over eighteen had children living at home with them. Of Quakers whose marriage date is known, 41 percent had been married one to nine years, another 39 percent ten to nineteen years, and 20 percent twenty years or more.

The life cycle can be reconstructed for only a few Quaker women who left records of their individual lives in the New Warke or Kennett Friends monthly meetings. The lives of over fifty women who married before 1751 can be partially re-created from these records. These women married on the average at twenty-two or twenty-three years of age, 70 percent of them to men a few years older than themselves; this pattern is similar to that found by Robert V. Wells in his study of Quaker marriage patterns. The women bore their first child at twenty-four and one thereafter every two and a half years for fifteen years. After age thirty-eight, having borne three boys and three girls, the women lived on with their husbands another thirteen years. They were as likely to survive their husbands as to be survived by them, and they died at nearly seventy years of age. Their lives were long and fertile, and much time and energy went into reproducing the next generation. That intensely active birth cycle was bounded at one end, in youth, by a long period of single life within the woman's own family and at the other, after childbearing, by another long period as a married or widowed woman with decreasing familial responsibilities (see Appendix table 2).[21]

Like other preindustrial rural women, New Warke women exhibited seasonality in their marriage and conception patterns, a variability also found in New England and in Europe during the late seventeenth and eighteenth centuries. The timing of marriages and conceptions seems to have been linked to the agricultural cycle in this rural community. Fewest couples married in May, the beginning of the summer fieldwork. Most married in August and September after the heaviest harvest work was done. Conception was least frequent in June, again a peak work month. It was most frequent in November, February, and March, times when women and men had few work responsibilities. Later, when Wilmington Quakers became more urbanized, they would spread marriages and conceptions more evenly through the year because work demands had little seasonality. Their favorite marriage month became June (see Appendix table 3).[22]

Whenever pregnancy occurred, it entailed some attention to clothing, of course. Simple gowns and petticoats could expand easily with pregnancy. Extra sewing for the children, perhaps less carefully done than for adult clothing, was necessary, but infants were probably swaddled, and then dressed in simple short gowns; boys did not wear breeches till age five or six. Thus the same infant clothes could be used for both sexes and passed down. The demands of pregnancy and lactation, then, beyond the need for more food for the mother and a few items of clothing, took little extra labor. Soon older children could begin to share child-care duties. Reproducing the farm labor force could not have been easy, yet it was not greatly disrupting, for it fit into the household cycle of production on the eighteenth century farm.[23]

As girls came of age, there was more work but more help, for daughters and daughters-in-law helped mothers when they remained nearby. When the reproduction cycle was over and grown children were able to assume duties, women had more time available for reading or writing, if they had learned to write, for nursing ailing friends and family, and for participating in the affairs of the church.

Representative of this average life, except for a somewhat later marriage and a life span of almost eighty years, was that of Elizabeth Reed Levis, a well-known Kennett minister. Born in the 1690s, Elizabeth married a man six years her senior in 1720. She bore her first daughter ten and a half months later and three more daughters and two sons in the next thirteen years, the last being born when she was forty. Her husband died of smallpox after twenty-seven years of marriage when she was in her early fifties. As a widow she saw her two sons and all but one daughter married. She died in 1775 as the American Revolution began.[24]

The New Warke Meeting, which was well supplied with an active elite of church women like Elizabeth Levis, had ten recognized women ministers. Lydia Dean died in 1750 but others would live on into the late eighteenth century. Thirty additional women, who appear in minutes of meetings in 1750, visited erring sisters, represented the meeting at quarterly and yearly meetings, and oversaw marriages.

The major community within which Elizabeth Levis lived out her life was that of the monthly meeting. From within this group of men and women, she drew her friends, received and offered daily support, met in women's monthly meetings, and received permission to travel. The meeting joined the rural women of the Brandywine Valley into one extended social unit, they looked after one another's welfare and social mores, visited and chastened wayward sisters when necessary, and formed a network of intricate intrafamilial ties often strengthened by intermarriage of children and business partnerships. The women of the meeting shared a worldview that made them different from the surrounding farm families, though they all shared a common regional and economic destiny.

Among the most important beliefs that made Quakers different was the great emphasis they placed on behavior as a reflection of spiritual health. As Barry Levy has pointed out, for eighteenth-century Quakers, right "conversation" meant right behavior. By the early eighteenth century, Quakers were in the process of becoming communities composed of individuals who inherited their right to membership at birth rather than becoming members voluntarily. Known as birthright Quakers, these young people were accepted by the meeting because their parents had exhibited proper "conversation," or behavior. The price parents paid for acceptance of their children into the religious community was responsibility for the children's actions. A child who did not repent of wrongdoing could be expelled. Moreover, a child who did not act in a proper way might also call into question the righteousness of the parent. Thus, the behavior of children as well as parents was important to maintaining the community.[25]

For most young people and their parents, a major test came as they emerged into adulthood and chose marriage partners with whom they would, in turn, help re-create the community. Because marriage was so crucial to the Quaker meetings, leaders devoted great time and attention to it. Over half the business in most meetings was apparently devoted to questions of marriage. The penalty for marrying without the consent of the meeting, or "out of unity" as it was called, could be disownment from the community. Whether disownment actually followed, however, depended on a wide range of personal and social conditions. Some people seemed more inclined than others to acknowledge that they had done wrong. Such public confession of wrongdoing and contrition usually led to the person's reacceptance. At other times, despite someone's reluctance to repent, the meeting felt more inclined to be lenient and the wayward could remain if he or she wished.[26]

By 1750, a marriage problem of crisis proportions had developed. There were actually two crises, similar to those that other religious communities experienced in the late eighteenth century. The first was a generational crisis in which parents lost control over the marital choices of their children. The second, arising from the first, was a marriage crisis in which Quakers were no longer able to ensure that youths who had grown up in the group married within it. The two were interrelated in a complex dialetic.

The late eighteenth-century generational crisis has been well documented. Young people exercised increasing autonomy in choosing marriage partners, seemingly because parents failed to provide a stable marriage system. Prenuptial conception may have been one strategy young people adopted to force parents to accept a marriage partner of their own choosing. It is difficult to trace prenuptial conception of those marrying out of unity, but there is evidence of increased premarital and extramarital sexual activity within the New Warke Meeting.[27]

Some of that activity can be traced in the rebukes issued to males in the meeting. S. F. was reprimanded for drinking, fighting, and keeping a woman as a wife to whom he was not married. W. C. rode with a woman and was observed to frequent the house where she dwelt "during her Residence there more than when She was not there, and he Might have avoided it." The women, of course, carried the greatest burden in the case of illegitimate births. H. P. "bore a bastard child." B. W., pregnant, hurriedly married out of unity. R. C. was also an unwed mother. Barry Levy found in a study of two other rural meetings that more young women than men married out in all classes and more disownments of women than men took place, especially among the poorer Quakers. As many as 50 percent of the poorer Quaker women married out and perhaps as many as 80 percent of these were disowned.[28]

Premarital conception seems to have increased during the late eighteenth century in all meetings in Chester County. Although, again, no records exist to document this precisely, the church notice of cases of premarital sexual relations, for women as well as men, had increased dramatically by 1770, as Jack D. Marietta has shown in his careful study of discipline in Pennsylvania meetings. Increased sexual activity does not seem to have involved married couples nor increased violence in sexual relations, such as rape. Young people, many of them later married out of unity, were simply determining for themselves whom they would marry without waiting for parental or community support.[29]

This generational crisis, in turn, provoked the marriage crisis within the Quaker meetings. In New England congregationalism, the reaction to the generational crisis was to open the church to new members brought in by young people through their exogamous marriages, but the response among Quakers was very different. Rather than welcoming or silently tolerating these marriages, Quakers forced young people to submit to their discipline by acknowledging publicly that they had erred. Those who refused could be disowned, which led to large numbers of disownments in the late eighteenth century.

Most of the young Pennsylvania women were disowned for prenuptial sexual relations or for marrying out of unity. Thus, the Quaker community lost its capacity to expand because of the insistence by parents that their children marry others of the proper conversation and legitimize the parent's own status within the community. Susan Forbes has shown that the New Garden Monthly Meeting, for example, made no disownments for religious differences or lack of moderation in speech or dress in the first fifty-seven years of its existence, from 1718 to 1774. But, meanwhile, after 1750, accusations skyrocketed with most disownments coming for marriage violations and fewer and fewer being pardoned. Marietta has estimated that after 1760 the Pennsylvania Society of Friends disowned over 20 percent of its members, a majority of them for violating marriage rules,[30] and that the community of Goshen probably lost a third of its membership through disownment between 1761 and 1775. In 1750, there were already some indications that New Warke adults could not maintain control over their young people. Of eight persons mentioned as marrying in late 1749 and 1750, over half married out of unity. One Kennett father, Alexander Fraizer, even provided in his will that daughter Miriam inherit part of his estate only "provided she behave well & marry among the people called Quakers."[31]

Marriage out of unity in New Warke was sometimes followed by reconciliation, as the women elders sought to obtain public confessions of repentance on the part of the young women. H. G. told visiting women, "I am sorry and desire to take Ye Blame on my Self." S. R. condemned her marriage procedure. S. K. acknowledged her marriage "hath brot sorrow & troble on my mind." A. M "condemned her disorder" and asked Friends "to pass it by." M. M. told the women, "I am sorry." B. W. acknowledged her marriage "contrary to the Sentiments of my own mind, which has a Long time been a trouble to me." H. P. asked her sisters to "bear with her a while." Others, however, remained unrepentant when Hannah Carleton, Mary Phillips, Rachel Pierce, Hannah Way, Martha Chandler, Dinah Gregg, or Esther Wilson visited them. Then women would ask the men to sign and read certificates of disownment at monthly meeting. Disownment may have indicated that the fabric of the religious community was unimportant to a woman. She may have already depended most on her family and the larger non-Quaker community for support, but disownment, symbolic expulsion, meant much to those still within the meeting. Even if the woman repented and remained within the meeting, children from an exogamous marriage were less welcome after 1762 when the Pennsylvania Yearly Meeting decided not to accept automatically these children's birthright. Endogamy gave one's children a securer place among the Friends.[32]

The disownments decimated the ranks of the Quaker women but left for those who remained a solid place in their communities and their families. Quaker women moved to the margins of secular culture, but there they exercised a greater amount of equality than did women who remained within the expanding evangelical and revivalist churches, such as the Presbyterians in Pennsylvania. The relationships among these women shaped the social world of the Friends, creating a dense web of friendships and relationships.[33]

A few migrants continued to come from Ireland and England but Quaker in-migration to the valley was mostly domestic, with families and individuals moving from more populous areas where land values were increasing. The Quakers carefully kept track of the origins of their members. The New Warke removal records for 1750 show that most households immigrated primarily from nearby areas of Chester County that had been settled in the first half of the century—New Garden, Concord, and Darby townships. Records also show that migration out had already begun to the even more sparsely populated southern frontier, principally North Carolina. The Quakers were a mobile population, moving southwest from areas close to Philadelphia but moving on, in turn, to places still farther south.[34]

These wandering Quakers went alone, in brother-sister pairs, as single parents, and in nuclear family groups. There was an excess of male over female emigration caused by marriages, since it was customary for the marriage to take place at the woman's meeting and then for the couple to remain there some months before removing to their new home. But females, like males, left and entered the community in search of better economic conditions. In the 1750s, women were an essential part of a mobile Quaker population moving to less populated regions with lower land values. By the 1800s, the direction was not south but farther west to Ohio, but the pattern had been established earlier.[35]

It is difficult to say what proportion of the population in the Brandywine Valley in 1750 was Quaker. Although neither Quaker nor tax records are complete for this early period, one can get some sense of Quaker numbers from comparing what records there are. Three of the earliest tax records available for the Brandywine Valley are one from 1737 for Brandywine Hundred, a second for 1745 from Christiana Hundred, and a third for Kennett Township for 1750. Quaker surnames account for 7 percent on the first tax list, 29 percent on the second, and 35 percent on the third. On later tax lists it is not easy to trace Quakers, but James T. Lemon has estimated that probably a third of the population of Chester County was still Quaker by 1790, even though in southeastern Pennsylvania as a whole the number had dropped below 10 percent.[36]

Even more revealing than the number of Quakers, however, was their place in the economic order. The Society of Friends had from its origins in seventeenth-century England attracted a relatively comfortable group of rural yeomen, well-to-do farmers who owned substantial amounts of land. For the most part, laborers and servants did not become Friends in large numbers, although producers and traders as well as artisans made up a sizable minority. This economic importance is already visible on the three early tax lists of 1737, 1745, and 1750. If one divides taxables from the three lists into quartiles, the Quakers appear in the first quartile greatly out of proportion to their numbers. In Christiana Hundred, for example, they make up 19 percent of the two lowest quartiles, but 35 percent of the two highest. In Kennett, they make up 11 percent of the lowest quartile and 26 percent of the second lowest, but 50 percent of the highest, and the second and third highest individual tax payers there were Quakers. In Christiana where the two highest tax payers were Quakers, two children of these families, Robert Richardson and Sarah Shipley, married in 1750. The economic visibility of Quakers becomes less clear after the mid-eighteenth century, however. A disproportionate number of the poor seem to have been people who were disowned during the generational and marriage crises. But despite this and the heavy burden of taxes for failure to participate in wars, the rural Quakers retained a comfortable financial niche by developing cohesive and productive family farms, making them the dominant culture in the region.[37]

Although Quaker males appear prominently on the tax rolls, women are also present. Almost no women appear on the 1735 list, but the 1745 Christiana Hundred list had 3 percent women and the 1750 Kennett list had 5 percent. Their number did not increase dramatically in the next thirty years, except in some townships as a result of the disruptions of war, but widows and unmarried women as well as wives played an important role as guardians of familial fortunes.

As the Brandywine Valley filled up with Quakers, it gradually assumed the characteristics of a Quaker culture area. Other English denominations peacefully coexisted, but the number of their churches did not appreciably increase between 1758 and 1800. By 1829 Chester County had thirty Friends meetinghouses. These rural Friends shared some of the characteristics of Philadelphia Quakers but their separate identity was sufficient to cause the majority of them to split from the more orthodox urban Quaker church in 1829. Within the larger socioreligious identity, rural Quakers were distinguished by their attention to the details of organized and regulated material life, their deemphasis of luxury and frills, and their industriousness, which impressed contemporaries. Thus the rural Quakers developed a culture that bound them together and provided leadership for the surrounding community. That leadership was always challenged in time of war but reasserted itself when peace was restored. The culture gave rural Quaker women a place that was unique in American culture, and the story of how these rural Quaker women interacted with other women and men of the area is one focus of this book.[38]

The following pages are not only about Quaker women; but because they emerged into public life early and maintained that leadership, they provided a large part of the private documents used to re-create the lives of rural women in the Brandywine Valley. The lives of the other women can be re-created, however, from other public documents. Their material lives, their education, their family composition, and particularly the butter making for which they became famous can all be described at some point during this hundred years of history. From less traditional sources, then, these women emerge to take their places beside their more visible Quaker neighbors.

The story is essentially one of how women responded to industrialization and urbanization by changing their work patterns. Each in her own way struggled against the bonds of family and community that held her in a subordinate place and, by doing so, loosened them. The mothers of these women had borne the rigors of settling a country and surviving its political and economic revolutions. Their daughters continued the pioneer tradition on other grounds. In the last analysis, these rural women were not able to create an agrarian vision that would unite rural ethnic and economic groups or broaden women's place in the developing rural capitalist economy. What a few of them did bequeath to other women, however, was an example of noncompliance with a system that demanded their submissiveness as well as their hard work. Moreover, in abolitionism and in the rural-based feminist movement of the 1840s, black and white women left an important legacy.

CHAPTER 2

Reproducing the Farm Family

On February 5, 1747, Sarah Heald Harlan of Kennett made her last will and testament, appointed her daughter Mary Evans executor, signed the document with a mark, and turned her mind to coming death. When she died less than a month later, Sarah Harlan left a substantial personal estate of 199 pounds. In addition to bonds and debts, she left almost one hundred yards of cloth, spinning equipment, sheep, cows, calves, heifers, pigs, and beehives to be divided among her grandchildren. Her married daughters, Charity, Elizabeth, and Mary were each to receive twenty pounds and she remembered a cousin Martha Way with four pounds. She had already settled her three surviving sons, George, Samuel, and Aaron, on her husband's plantations as he had requested when he died fifteen years earlier. She had performed her earthly work well if the care with which she tended her family's concerns are a measure. She had reproduced the farm family. Her work was done.[1]

In many ways, Sarah Heald Harlan of Kennett was almost a prototypical middling farm woman of the early eighteenth century. Born in 1692 in Cheshire, England, she migrated to the Brandywine Valley with her parents when she was ten. Sarah came of age within a traditional patriarchal family structure, one in which the power and control by the husband and father was codified and enforced by an external male-dominated legal system. A wife could exercise power only as a surrogate for her husband. Age fourteen probably brought the first lessening of parental authority for Sarah, for if orphaned, a child could choose her own guardian at that age. Until she was eighteen, however, she owed her labor to her parents or her guardian. From age eighteen to marriage,

Sarah probably continued to work for her parents, although she now spent at least part of her time processing fibers, spinning yarn, occasionally weaving textiles, and sewing linens for her future household. Young women who worked for others while living at home seem to have turned money or payment in kind over to the father. Probably Sarah did not work outside her farm household, for there was enough to keep her busy there. A legacy might have been Sarah's at eighteen had her father died earlier. Because he did not, she probably received furniture or a sum of money upon her marriage.

Sarah married Aaron Harlan in 1714, when she was twenty-one and he was twenty-nine. Aaron was a first-generation immigrant, born in Ulster, Ireland, and brought to the Brandywine Valley when he was two by his mother Elizabeth Duck Harlan and his father George Harlan, who were among the first Quaker immigrants in 1687. Neither Elizabeth nor George Harlan lived to see their son Aaron marry, but despite his having eight brothers and sisters, the parents had given him land.

Marriage allowed Sarah the freedom of her own household but no legal control. At marriage, any property she owned became legally her husband's; anything she inherited became his; and any income she brought into the household was also his. She bore seven children in the eighteen years before Aaron died in 1732. The four sons, George, Samuel, Aaron, and Jacob, were his male heirs.

At his death, Aaron divided his 530-acre plantation on the Brandywine between his sons George and Samuel, set aside a 470-acre plantation in Kennett for young Aaron, and provided that Jacob receive an equal portion. Sarah had the task of carrying out this division of the patrimony. She also arranged, under an agreement made with her husband, to pay her father and mother, who lived on young Aaron's Kennett plantation until he came of age, from the estate. In order to support herself and raise their children, she received one-third of her husband's real estate for life or as long as she remained a widow, and his personal estate. She accomplished all this although son Jacob died in the interval. She hired harvest crews and kept the plantation running. She not only reproduced the farm family; she also did it within the guidelines of eighteenth-century patriarchy and yeoman mercantile capitalism.

Sarah Harlan's life cycle fit neatly into three almost equal parts. During an apprenticeship of about twenty years, she learned from her mother how to care for a family's physical needs, feeding, clothing, and nursing them. During eighteen years of married life, she bore and raised seven children, and managed the farm household, perhaps including a large amount of textile preparation. Then, in her widowhood, she took over complete management of the plantation until she could pass it on to her sons.

In all this, there was little indication that Sarah Harlan considered herself as other than an agent for the man she had married and the economic and familial system already established. Only two things seem uncommon, that she named her daughter as executor rather than a son or male relative and that she remembered with a small bequest a young female cousin not yet of age. These were ways in which Sarah Harlan could express the separate interests that women might have in a system they were so crucial in perpetuating. There were not many.

Sarah Harlan is thus a good woman with whom to begin this analysis. Her fertility, her service to the family, her maintenance of the core nuclear family structure, and her passing on to sons the property that would allow them to maintain their patriarchal role in the family constituted the main elements in women's reproduction of the American farm family. The first of these, fertility, was central to the farm family's intergenerational survival. Children provided labor for the farm before marriage, established ties through marriage to other families that would enhance the family's chance for economic stability and social status, and later assured care for parents and often siblings when they became unable to care for themselves. Sarah's seven children, then, were her contribution to this process. The naming of two of Sarah's children after her mother and father-in-law and one after her husband signified her commitment to carrying on not only the patrilineal but also the matrilineal names.

Service was Sarah's second crucial contribution. For fertility to count, women had to see that their children were prepared physically and socially to assume their responsibilities in the next cycle of reproduction. This involved physical care for each child, and the skill with which this was done would help determine how many children survived to reproduce in their turn. That six of Sarah's seven children survived to marry and reproduce was due to a combination of factors: their rural isolation from communicable diseases, her personal care, and the availability of a varied diet for herself and her children that included ample quantities of pork, beef, milk, and honey.

Socialization is more difficult to identify, but presumably the Harlans' six children all received the proper gender-based skills to fit them for rural life. The males were probably educated to write by a male relative or hired scholar or by an older brother, for Sarah could not teach them herself. But of the practical day-to-day operation of the plantation she had much to teach both sons and daughters. She had no bound servant to help her with her chores, although a male servant helped her husband at his tasks. She might have had the help of hired girl or tenant woman in producing textiles, for the farm had thirty-nine sheep at the time Aaron died. Even three daughters would have difficulty processing that much wool into yarn. The presence of a large quantity of yardage at her husband's death indicates that at least some of the wool and flax was processed at home even if the actual weaving was perhaps done elsewhere. The household itself did not require much servicing, for it consisted only of four chairs, a table, four beds, and a chest of drawers, some sheets, pillows, tablecloths, napkins, pewter, and brass. Kitchen equipment was equally simple—iron pots and a pot rack, a frying pan, gridiron, and tongs. She had three cows to milk but no churn or cheese press. Thus household life, except for textile processing, was simple. The family produced and processed mainly for its own use and some surplus to sell on the local market.

Passing on the land to sons and cash to daughters was Sarah's most crucial role in reproducing the economic structure. As executor of her husband's estate, she participated in passing on both her own and his property. The land that the sons inherited, however, was greatly out of proportion to what the daughters received. The Kennett plantation her son Aaron inherited was worth 450 pounds, whereas her daughter Mary received twenty pounds in cash. Both received equal shares of the remainder of her estate.

Had the land that Sarah inherited been encumbered by her husband's debts, she would have had to sell it to satisfy creditors under a 1688 Pennsylvania law. Widows in Bucks County in the second half of the eighteenth century, for example, sold family farms to meet their husband's debts. Fortunately, the Harlan estate was in no danger. But the legal climate of Pennsylvania, as Marylynn Salmon has observed, was not liberal for married women—or even for widows who struggled to pass property on to their children.[2] Thus the power Sarah held as widow and executor was not passed on to her daughter, even though Mary in turn became her mother's executor. Any property Mary inherited was legally controlled by her own husband unless her mother made special arrangements. This normally did not occur unless there was some particular reason for the parents to distrust the son-in-law. With rare exceptions, the son-in-law was accepted as a relative who did not receive land but who was expected to receive some cash from the parent's estate and, in turn, had to provide a third of his estate for life or widowhood just as Aaron had done for Sarah. Everything they shared as husband and wife belonged to him, including the bed they slept on, the pots she cooked in, and the horse and sidesaddle she rode on. Only her clothes were her personal estate.

Like other rural women of the time, Sarah moved within the boundaries of both family and culture. Property passed from generation to generation according to rules sanctioned by the community and formally regulated by family law. Because the law stipulated that patterns different from the norm be documented in wills, the shadow of a woman's pattern distinctly different from that of men can be found there. Systematic differences emerge between men and women, differences that clearly reflect a distinctive relationship to property for women. Wills reflect inheritance strategies through which families reproduced themselves—passed on power and wealth—and the role women played in that reproduction.

Eighteenth-century American family laws did not differ significantly from English common law. Both were based on the doctrine of marital unity that obliterated a wife's legal identity and mandated her subservience to her husband. In return for the loss of her civil rights, the woman was assured no less than one-third of the husband's estate when he died. Real property was not, however, hers to dispose of or, often, even to manage. Men's wills settled distribution of land among children; widows were bound by the constraints of these wills and often only had personal property to dispose of as they wished. Some eighteenth-century wills deprived her of even maintenance if she remarried, in order to keep the husband's property from falling into the hands of a new male line. In practice, equity laws allowed some economic autonomy through prenuptial and postnuptial contracts and trusts, but as legal historians have concluded, equity gave control over property to only a few women who had large estates and legal advice. There is no clear evidence that the American practice was any more favorable to women than the English practice, though both may have deviated in some procedural ways from the theoretical norm of marital unity.[3]

Constituting the basic economic unit of society, family members did not normally engage in economic exchange with one another. Outside the household men traded for profit, subject only to the confines of British mercantile and common law. Within the household, they shared resources. Through marriage, as Miranda Chaytor has pointed out, kin groups shared and redistributed resources in households. But as Chaytor also noted, because gender relations were asymmetrical, sharing was also asymmetrical. Although women were often active economic agents within marriage, they and their families gave up any claim to the husband's property in return for her maintenance if he died, unless they agreed before the marriage on some other arrangement, special circumstances caused the courts to intervene, or the husband himself explicitly ordered, usually in writing, some other arrangement. The wife could, and in practice often did, act as an agent of the husband during his temporary absence or death. This role was very important in many rural families. It meant that a woman could exercise great power as her husband's surrogate. She could not exercise that power in her own right but neither could she be ignored. The younger generation could not acquire complete property rights until she died.[4]

The daughter occupied a very different position than the wife. In English and American society, inheritance was bilateral, with estates being divided among sons and daughters. Here, again, the norm was one of asymmetry: sons, especially the oldest, normally received more than the daughters, and they received land while daughters received cash or other valuables. If already married, of course, a daughter's property passed legally to the husband; if single, a woman might control the wealth herself, although it might be put in trust for her. Still, this right to part of the patrimony gave her some potential power within the family. Jack Goody estimates that in Europe roughly 20 percent of all families would have only daughters to inherit their estates and that therefore daughters had great quantities of land under direct or indirect control. Thus, by the eighteenth century, female ownership had become very important.[5]

In England, where the principle of sharing existed, there was always tension between the desire to maintain family land intact and yet to provide for all the children. The existence of large amounts of available land in the colonies made it possible to provide generously for more than one heir, thus lessening these tensions for the first few generations. By the mid-eighteenth century, however, these tensions were mounting again with the increase in land values and the distance children had to travel to claim new land. Since the status of rural households and individuals within those households still depended mostly on the amount of land they controlled, shortage of land was bound to increase tensions.

Families proceed through stages, making the question of inheritance still more complex. The first stage brings the property of two individuals together. Births give children a claim to that property. The fission of the family when the children reach adulthood, marry, and often receive part of their inheritance brings yet a third stage, and the death of the parents brings the final partition of family property.[6]

The life of Sarah Harlan is a good example of how these stages operated in the Brandywine Valley. It is possible, however, to go beyond individual cases like Sarah's to examine the lives of a larger number of women. By using the written remnants of a few rural women's lives, one can reconstruct the lives of a larger group that might show other patterns. These can also be contrasted with women's role in reproducing the rural family in 1790 to see what differences the lives of Sarah and her daughter might exhibit.

Pennsylvania laws, however, had some variations from these generalizations that should be noted. During the period from 1750 to 1820, dower laws were gradually expanded to provide women more rights to family property. Dower remained one-third of the husband's real property for life and one-third of his personal property forever if there were children. If there were no heirs, the wife received her husband's entire estate. Simple agreements between women and men could provide women with separate estates. But Pennsylvania had no separate court of equity, separate trusts for women appear to have been infrequent, and the right of women to devise their own property if they had a prenuptial agreement but no trustees was challenged as late at 1793. Husbands continued to have extensive managerial power over wives' real property, even though a husband could not convey title of property to someone else without his wife's consent. Although women's dower rights were strictly enforced—one Pennsylvania Supreme Court justice even referred to "life, liberty, and dower"—they were badly eroded by a law that allowed a husband's debts to be paid before a widow received her portion. Before 1782, if a husband died without a will, Pennsylvania law gave the widow one-third and divided the remainder among the children unequally, giving the eldest son a double portion. After 1782, all children received an equal portion. The will was an important way to make special provisions that differed from this standard.[7]

When one compares the 6 percent of Chester County wills left by women in the 1750s with that of Sarah Harlan, it is clear that they form a pattern. As was common in western bilateral kinship systems wherein women were responsible for maintaining kin relationships, women were bound by the constraints of their husband's wills to leave estates to sons or to male kin. But commonly, a large number of daughters also inherited land. In one case, a daughter who received a 150-acre estate from her mother had no brothers. Another daughter received an unspecified amount of land. Yet another daughter received the remainder of an estate to the exclusion of her brother, who received a large amount of cash—three hundred pounds. Two others received the remainder of the mother's estate equally with their brothers. Most mothers, however, left only personal property to their daughters, usually clothing and household goods. Fifteen percent left the whole or an equal portion of their estates to daughters. Nuclear family kin, then, sons but also daughters, received most property (see Appendix table 4).[8]

Two women made important stipulations in their wills that indicate what may have been a more common concern. Both specified that the son-in-law not be allowed to completely control property inherited by a married daughter from her mother. Mary Waln of Norrington, for example, left one-third of her estate to her son Jonathan Lewis, and one-third to each of her two daughters Hannah and Ann, excluding their husbands from all control and giving the daughters the right to dispose of the property as they wished. Mary Hayward of Concord gave property in trust to her daughter Martha with explicit instructions that husband James Wilson "shall not in ye Least Intermedle." If James died, Martha was to receive the money; if she died, the money was to go to her children equally.

By such arrangements, mothers could ensure that their daughters controlled their inheritance. Most, however, trusted sons. Four also made sons-in-law coexecutors, and one, sole executor where the estate was left to her daughter. No daughters were made sole executors and only three were coexecutors with other males. Daughters-in-law might be "loving" but received personal property only. In the 1790s, women seem to have increased their use of male friends to help with estates, but they also used sons-in-law, perhaps indicating a concern that daughters and their children receive their due from brothers (see Appendix table 5).

Although women had little freedom to dispose of the bulk of family property, they had considerable freedom in transmitting small amounts of money and property. Women most often left bequests to near-kin, nuclear kin, and non-kin in that order. Granddaughters often received money gifts and one received half of an estate, although none received a two-hundred acre plantation such as Jane Beverly willed to her grandson. Ann Pyle left her sister, and Isabella Mitchell her niece, the remainder of their estates. Barbara Wilson willed her mother a house and lot. Of non-kin bequests, the most important was that of Margaret Ellis of Haverford who left her friend Mary Kirk one half of her estate equally with a brother and made them joint executors.

Women increased the number of bequests to nuclear and near-kin in the 1790s with children and grandchildren clearly the gainers. But women expressed even more clearly their appreciation of the support of female friends. Alice Sarah Gilbert of Westtown left her estate to friend Susannah Hoopes, "to her alone in grateful Remembrance during my Weakness & Infirmity & as a Token of my grateful Acknowledgement thereof." Rachel Stevens of Uwchlan left the remainder of her estate "in consideration of the trouble I have been to my dear friend Susanah Evans."

In contrast to women, males kept their inheritance closer to home, giving most to nuclear and near-kin and far less to distant kin. Most males left land to their sons along with varying amounts of money. There was little favoritism of older sons explicit in husbands' wills, for only Henry Caldwell of Newtown signified a double portion for his eldest son. Since most sons seem to have already received property from living parents, it is difficult to compare the bequests of sons and daughters, but only when there were no sons did daughters receive land. Daughters normally received personal effects or sums ranging from five pounds to one hundred pounds. Isaac Vernon of West Bradford left his two sons lots and his two daughters equal shares of the remainder with his wife; John Hamilton divided his estate equally among sons and daughters. Near-kin—grandchildren, brothers, sisters, nephews and mothers—received land or money infrequently. Joseph Wilson of West Fallowfield left money solely for his daughter and excluded his son-in-law, but the concern was not expressed in so extreme a form as was that of the mothers' mentioned earlier. John Allen of East Marlborough, who gave sums to a brother, sister, and mother, had neither wife nor children to inherit.

Women did not leave a large amount of money to religious institutions. Only one left money to the church, a woman who explicitly left her estate to the Baptist poor after bequeathing a brother and sister only five shillings each. Men endowed churches more frequently than did women. John Allen left twenty pounds to two Quaker meetings; two other men left eight pounds and six pounds. Males remained the largest givers to charity into the 1790s and were also able to leave larger amounts of money to friends. Sebastian Root of Coventry left one hundred pounds to charity and three hundred pounds to a friend (see Appendix table 4).[9]

Few wives received anything but personal property—usually their own, but sometimes a part of their husbands' personal estate. In almost every case, wives received land or the use of a house only for their widowhood and usually to care for underage children. Even in special wills, a widow usually only received one-third for life. Husbands cared for their wives in traditional ways. Seldom did a husband make his wife sole executor, although he often made her executor with a brother or son. Zachariah Butcher of East Nottingham chose a daughter as coexecutor, but he had no sons, only seven daughters. These men were very traditional, even when they deviated from the norm in making wills. In the 1790s, men still most often appointed sons and male friends as sole executors (see Appendix table 5).

Females showed clear differences from men in passing on their wealth. As late as the 1790s, they still dispersed wealth over a far wider range than men. They gave less to nuclear kin and far more to near-kin, especially grandchildren, and to distant kin, especially nieces and nephews. The pattern that emerges most clearly is a female concern for grandchildren and a greater tendency to leave money to their own sisters and brothers. It should be remembered, however, that none of the women had husbands to whom they could leave an estate. This meant that when women did have estates to disperse, they were likely to be benefactors of a broader constituency than men simply because they had no husbands. This allowed them a wider dispersal network and gave them a special place in the intergenerational transfer of wealth (see Appendix table 4).[10]

Chester County males, as late as the 1790s, showed little inclination to broaden their wives' economic role in the transfer to wealth. Most dying men trusted their wives to bring up their underage children, but few deviated from the standard one-third of the estate for life for the widow. Careful provisions for children do indicate trust in wives assuming this responsibility legally. John Varley of Conventry left his plantation to his wife Magdalene Varley for six years to bring up and educate their three children. However, only one-fifth of the departing fathers trusted the mothers with estates and the wife often shared this responsibility with a son, whereas a third of the sons handled the estates themselves. This was a task to be performed primarily for underage children with control by husbands' friends or sons as executors. At the end of the six years specified in the Varley will, for example, his executors, two male friends, were to sell the estate and divide the money among Magdalene and her seven children. Sole executors were most common in men's wills, which frequently appointed sons, wives, and friends. Mothers did not increase their reliance on sons as much as did fathers. The greatest increase for men was sons and male friends. There was only a slight increase in female power but mainly a shifting to new males. Patriarchy was not in disarray, but it was certainly changing. The locus of power in the family was now split between males, with women exercising more power through males from outside the family as well as slightly more direct power. Men, on the other hand, increased their dependence on sons as well as on friends to a greater degree than their dependence on women (see Appendix table 5).

Although transfer of property was important, ultimate success in reproducing the farm family meant success in reproducing only enough children for the available land. The problem for landed women was to adjust their fertility to reproduce enough children to survive to inherit the family wealth but not so many that the wealth was not sufficient to maintain the family standard of living. Land in the Brandywine Valley was still available for Sarah Harlan's children, but her grandchildren were already experiencing the pinch of increasing population and land values. Quaker family strategy dictated a decrease in fertility as land became less available. Women could accomplish this in a number of ways: remaining single, marrying later, lowering the age at which they completed their family, or using some artificial means to limit births. The first two were the earliest to be employed by both Quaker and other communities. The efforts to limit children resulted in what historians often label the "demographic transition," a decline in fertility rates in rural areas that began in the late eighteenth century.[11]

The strategy of not marrying at all was one option for Quaker farm daughters. There is some evidence that this option was utilized more frequently after the 1780s. In one study, Robert V. Wells has shown a decrease in the number of daughters marrying to less than 50 percent and an increase to almost 25 percent of those living to age fifty remaining single. The decision not to marry would have had an important impact on farm families. Given the lack of alternatives for these single women, they may have influenced the gradual restructuring of families. Staying at home longer or joining families of kin as they aged, these women might have moved families from a nuclear to an extended structure. The presence of unmarried daughters was a new element in rural families and one that eventually may have become a crucial issue because of the increased demands for their maintenance.[12]

Even when Quaker women did marry, their fertility rates had declined by the revolutionary era. Wells has shown that Quaker women's family size declined from almost seven children before the Revolution to fewer than six children after. Other historians have found similarly small family size for Quaker families in late eighteenth-century Chester County.[13]

During much of the eighteenth century, delay of marriage rather than not marrying seems to have been the preferred strategy to achieve reductions in fertility. At least one part of the decline in fertility can be accounted for by later marriages. Wells estimates that Quaker women born by 1730 married at 22, whereas those born between 1756 and 1785 married at 23.4. Similar shifts took place in some New England communities, too.[14]

Late marriage was not the only or even the main strategy, however. Wells could account for about only a fifth of the decline in fertility by raising the age of marriage. The major reason seems to have been a reduction in the level of childbearing within marriage. By 1756 to 1785, women not only married later but also deliberately limited the size of their families by having their last children at an age earlier than had prerevolutionary women and by increasing the birth intervals of their last children. By the end of the eighteenth century, Quaker women were reproducing well below other American women.[15]

The intent to reduce fertility and the resulting decrease seem clear. The method is not. Men and women most likely cooperated in using abstention, especially in the later stages of marriage. Abstinence may have been thought of as an asceticism within marriage or simply as a new thriftiness in sexual activity. The decision to abstain may also have brought a new level of cooperation to Quaker marriages, something difficult to document but perhaps signaled by a greater emphasis on the importance of compatibility in the late eighteenth century. What Brissot de Warville noted among Quakers as the "bonds of family and loyalty" may still have been patriarchal in form but based on a new need to change the content of marriage, making it more cooperative.[16]

On the other hand, there were techniques that both the wife and the husband could use independently. The use of coitus interruptus was one method males may have used within marriage. The probable spread of this technique in the eighteenth century is still something of a mystery to historians. Jean-Louis Flaudrin has argued that in France, such practices were first used outside of, then within marriage. Although other historians have questioned his explanation, there is little other evidence to show how this contraceptive tradition might have been learned and transmitted. It was used, according to Lawrence Stone, in late eighteenth-century Massachusetts. The practice was well enough known that the utopian community of Oneida used it as the basis for their complex marriages in the early nineteenth century. Therefore, the practice must have been well known among some groups of men by the late eighteenth century.[17]

Wives also exercised greater responsibility in preventing conception. Length of lactation was one of the best-known means of delaying conception. Proverbs often encouraged lengthened time of breast-feeding to avoid conception, and scholars have confirmed that lactation does reduce fertility, especially in the aggregate. In some cultures, it is common for a woman to breast-feed her last child for as long as five years. It is possible that Quaker women extended breast-feeding in the late 1700s, and the practice might also have been related functionally to the increased emphasis on child-centered Quaker families.[18]

There was one final practice that at least some Quaker women may have used to reduce fertility. They may have been among the first Americans to use what might be termed artificial birth control in their search for smaller families. In discussing sexual customs in the 1790s, Moreau de St. Méry noted that Philadelphia apothecaries carried syringes, first imported by the French and then adopted by the Quakers. Although the decline in fertility was already in progress by the end of the eighteenth century, the search for aids to further ease the burden of married couples already committed to fewer children may have led to the use of the syringe for douching. Medical authorities are still not in agreement on how much douching can reduce fertility, but like breast-feeding, douching can probably affect the aggregate fertility rates in a community. Moreover, when couples use more than one method of birth control, the chances of controlling conception increase. Thus, the availability of several alternatives may have been a factor in reducing fertility within marriage.[19]

Household censuses taken by Chester County and New Castle County officials in 1782 and 1783 also allow a comparison with Quaker families. The average household size of six Chester County townships in 1783 was 6.7, larger than the Quaker family size of 6.2. However, these were households, not nuclear families, and included servants. In Brandywine and Christiana hundreds, where servants were not included in the census, the family size was much smaller, less than 5 per household. The evidence thus points to a decline in fertility toward the end of the century for all families not just Quakers. It is possible that male deaths from war would have affected non-Quaker families more than pacifist Quakers who kept their sons at home or that younger members were absent working and living on other farms. Still, in Brandywine Hundred, where only 2 percent of the households were headed by females, the household size was only slightly over 5. Nuclear farm families in Brandywine Hundred almost sixty-five years later were only slightly smaller, indicating a steady drop among the population at large (see Appendix tables 6, 7, and 8).

Reductions in fertility meant that the women had to provide better health care for themselves and their children so that this smaller number of children, in turn, could reproduce. Women were the primary health care providers everywhere, but their responsibilities were especially heavy in rural areas such as Chester County.

In the late eighteenth century many rural women depended on lay doctors—male and female—for assistance in birthing and caring for their children. The tradition was later recalled by William Darlington of Birmingham who, although he was a trained physician, said he had been accepted by the country people because they had been given medical assistance by his great-grandfather Abraham Darlington and Abraham's children Thomas and Rachel. Unfortunately, little is known of Rachel Darlington's medical practice or of her father Abraham's care of female patients. Abraham divided his books on "physic and chirurgery" equally between his son and daughter at his death in 1776. At that time Rachel Darlington Seal was a widow practicing medicine in Birmingham. Beyond that nothing is known of her work. It seems likely that no female descendants carried on Rachel Darlington's work directly, for by the 1770s, William Shippen had already mounted his antimidwife campaign in Philadelphia and male midwives were gaining ground while the number of female midwives was declining in proportion to males.[20]

William Darlington, on the other hand, received his medical training from a Wilmington doctor and returned to the family farm in Birmingham to practice medicine. A restless young physician who eventually became wealthy enough to discontinue his practice, he left us, along with his discontented musings, a unique health record of the women of Chester County. Because he ministered to the wealthy and the poor, black and white, young and old, his records provide a glimpse into family health and into one of the central tasks of women—reproducing and caring for the farm family. Between 1804 and 1812, he kept a log of 109 births, which together with his other diaries and accounts allows a fairly complete documentation of the health cycle of one group of rural Chester County women.[21]

Only slightly more girls than boys were born (52 percent). At birth females held no significant advantage over males, although they had fewer birth defects and fewer birth problems. Seven percent of all babies died at or before birth and another 3 percent had serious birth defects, whereas 13 percent of male babies died or had birth defects. The deaths were due to various causes, some of which are not clear from the records. One died at five months because of uterine hemorrhage; a second died two weeks before birth; a third died in the womb at seven and a half months. One female and one male fetus literally bled to death. "The bed was deluged in blood, which had run through it and on the floor," Darlington recorded in one case. One female child was born the size of a "rat" and died within twenty-four hours; the second was born "weakly."

Some birth defects were recorded among the babies who were born alive. One was deaf and mute, another had a harelip, a third a tumor in the scrotum that the doctor did not consider serious. The fourth, the only female, had a "warty excressence" on her head. Although physically disfiguring, none seem to have posed particular problems in a rural setting. One child had to be resuscitated at birth; after being put in water, chafed, and having his lungs blown into by the doctor, he apparently revived and became a normal baby.

Darlington reported surprisingly few infant deaths. There were other doctors in the area, at least judging from his reports, and infant deaths may have occurred swiftly before a doctor could be called. Nevertheless, there is little discussion of infant illnesses in Darlington's records. Urban doctors who believed that lower infant mortality existed in the country because it was cleaner and fresher were probably right.[22]

Childhood illness was frequent but seldom fatal. At this age, girls shared with boys a range of illnesses needing more care than mothers' home remedies. They contracted whooping cough, colic, coughs, fevers, diarrhea, and rashes from poisonous plants. Darlington reported few accidents—a scald, a dog bite, a severe fall—and measles, too, were infrequent. Cholera appeared once but was not widespread, and a number of parents had Darlington vaccinate their children. Only one female child had a serious problem—a tumor in the groin—and at least one child died of convulsions apparently brought on by a bad case of worms.

Young women appear to have been healthy as well. One young black woman, Sal, died. The young doctor was never sure of the cause, although she had symptoms of typhus, and her body was infested with worms. The usual remedies of bleeding and dosing with salts and opium did not help. A few patients suffered fevers and chills; one, a "robust woman," incurred a chill by washing yarn in the Brandywine in March. One woman had apoplexy, perhaps a symptom of epilepsy, but it does not appear to have recurred. The doctor mentioned no menstrual problems among young single women perhaps because the average age of menarche among Euro-Americans at this time may have been just over fifteen.

Pregnancy did not usually occur until after the age of seventeen. Darlington considered even this "quite young" for childbirth, for he recorded assisting a woman of this age at birth with a special comment. Most of the pregnancies seem to have been normal, both socially and physically. Ninety-four percent of women were married at the time of birth. In fact, these women were identified only as the wives of males whose names Darlington meticulously recorded. Prenatal care was the responsibility of the woman herself unless she had some complicating illness. One woman mentioned she always had varicose veins when she became pregnant—she could tell when she was by the swelling. One had periodic headaches during the day, others dyspepsia. Darlington prescribed the usual bleeding and dosing with salts and calomel. One woman miscarried after contracting a severe chill. If others miscarried, this doctor was not called.

What are now euphemistically called problem pregnancies are mentioned for the first time in Darlington's diary after he moved to the village of West Chester in 1809. About 6 percent of all births were to unwed mothers. Mary Montgomery, Mary Darlington, Suzy Evanson, Phebe Jacobs, Lizzy Temple, Zillah Lewis, and Hannah Dilworth all bore children without the support of fathers or the sanction of society. Of these seven, at least two were listed by Darlington as "the unfortunate victim of Seduction," though whether this was physical or psychological coercion is not clear. Two others the doctor listed without sympathy, one who had a child "which some Irishman had helped her to," and another who "contrived to get herself with child." Two others seem to have incurred more blame from the young doctor—one pregnant from "too great proximity to the rougher sex" and another "a sort of widow bewitched, who fell in Labor in consequence of too much familiarity with one of the rougher sex." Comments like these indicate that perhaps a third of such pregnancies were forced on women, a third were the result of consent, and a third occurred among women who had a more active sex life than the customs of the time condoned. Such a small sample and Darlington's brief comments cannot do more than suggest the varieties of experience of these unwed mothers.[23]

Childbirth seems to have been safe during this period. No woman died giving birth, though 7 percent of the infants did, often after painful labor. Almost half the women had an easy labor, several giving birth within a half hour of the first pains. In one case, Darlington left to perform a quick errand and the baby was born before he returned. Another 30 percent, however, had either a "severe" or a "tedious" labor—difficult, long, or both. Four percent suffered a breech birth or some other difficulty. The doctor had to rupture the membranes of two women with a knitting needle or pin.

Only one woman had a physical problem that prevented a live birth. This was the sort of case that women, midwives, and doctors feared. Darlington was called in for consultation with another doctor when, after twenty-four hours of painful labor a young woman could not deliver because of a deformed pelvis that made it impossible for the child's head to pass through. It was her second and presumably the second time the baby had to be "delivered by the crochet," a method whereby the doctors opened the baby's head and delivered it dead. In this case the infant was a fine male, "a perplexing and distressing scene," Darlington reported.[24]

Eli Rathew's wife, as Darlington listed her, was not the only woman in Chester County to have such a difficulty. But, apparently, two older physicians rather than the young Darlington handled difficult births of this type. A second young doctor, a recent graduate of the University of Philadelphia, was also called in for difficult labors to deliver the children with the help of forceps. A later account of one birth reported that women objecting to the old method, insisted that a doctor trained to use forceps be called to attempt a live delivery. Midwives seem to have refused to use forceps feeling they were a dangerous substitute for their hands. Because of this, they gradually lost their practice. A doctor, once called in on an obstetrics case, was usually retained as the family doctor and thus found obstetrics a good way to expand his practice. Thus doctors actively competed with midwives for childbirth cases. In Chester County, women seem to have sought out doctors rather than midwives by the turn of the century.[25]

Labor and delivery practices are not well documented. Doctors of the time thought exercise and a light diet could help ease labor in normal births. They also thought countrywomen had an easier time of it than urban women because of their active lives; some doctors urged urban women to exercise more. Darlington recorded a variety of birthing methods. Sometimes, he said, he gave laudanum to increase the frequency of birth pains; other times he might have a woman walk about for this purpose. One woman gave birth sitting on the side of the bed, and in another case, the doctor elevated the hips to make birth easier. One black woman, Darlington reported, "speedily delivered as she sat on another woman's lap." Thus, in the nineteenth century, birthing techniques did not yet necessarily require the woman to lie prone.[26]

West Chester women probably had extensive help from relatives and experienced friends during and following childbirth. It is likely that the presence of doctors did not mean that women preferred them but rather saw them as a supplement to other women's assistance. Standard early nineteenth-century practice was for the new mother to spend one week in bed and then several more resting and sitting up in bed. For farm women, who had heavy duties and were in generally good health, the resting period may have been brief. A few women experienced postpartum difficulties, suffering afterpains or hemorrhages. Only one had puerperal fever, a rarity in the area in the first decade but more common after 1812. Probably most painful was mastitis, which several women developed while lactating. The breast in that case might have to be lanced. Still, only two of the new mothers had such difficulty. Birth-related medical problems seem to have been unusual. Most mothers were healthy during and after childbirth.[27]

As adults, however, women had gender-related complaints. Most frequent was what doctors called "suppressed menses," a term that covered everything from undetected pregnancy to serious disorders. Aside from these problems, women were quite healthy, although they suffered from occasional pains, rheumatism, hepatitis, digestive problems, hemorrhoids, fevers, skin inflammations, headaches, falls from horses, and toothache from "one or two old stumps of teeth." One was an alcoholic.[28]

As a general practitioner, Darlington also acted as psychiatrist as well as dentist and doctor. He treated two female patients for mental problems, one of which he listed as "hysterical." The other case was more complex. John Davis's wife, four months after the birth of her child, was not well, her ideas being "deranged." She complained of pains in her head, "her eyes rolled with all the wildness of a maniac—sighed deeply—& talked strangely." In this case, the husband dismissed the doctor, saying "nothing but Devilishness ails her." Here was certainly not a case of labeling deviance as insanity but rather dismissing severe symptoms of physical or mental disorder as self-induced. One wonders if, in other times, witchcraft accusations might have followed. As for Davis, she simply never appeared in Darlington's records again.[29]

Old women seem to have infrequently sought Darlington's services. He listed old Sarah Dilworth as "being too crazy" to pay a twenty-five-cent fee for treatment of an infection. For others, he treated physical symptoms of old age—rheumatism, cataracts, flatulence, skin troubles, kidney stones. In one case, doctors removed a cancerous breast from an old woman in the poorhouse; she died. But not many old women were even identified as such.[30]

Medically, men differed from women in some important respects. Doctors were called more often for male work-related accidents—a harvest sickle cut, a hand mangled in a mill. A few were injured in drunken brawls. Several had syphilis or gonorrhea. Men seem to have had more flus and fevers than women.[31]

On the whole, females, even with pregnancy and gynecological problems, saw doctors less than did males. For a two-year period, Darlington kept a record of health problems by gender. He finally gave up the listing because it was too difficult to separate the illnesses, but the gender count was accurate. In a twenty-four-month period, 54 percent of his patients were male. It is possible that because women might have been able to continue domestic work in spite of illness, they did not resort to doctors as frequently as males. On the other hand, by the early nineteenth century, females may have been exhibiting the physical characteristics that would make them the stronger sex in terms of overall health. Despite the use of bleeding—so frequent that one male "fainted at the thought of it"—in amounts of up to twenty ounces and the use of opium in large doses, rural women survived the intervention of doctors and successfully cared for children and themselves.[32]

Thus, women through their caretaking performed an essential function in reproducing the farm family. During this time, they seemed to have performed the task more and more effectively, birthing healthy children, maintaining their own health, and caring for children until they could inherit the wealth collected by the family. Although essential, women's work did not materially change the patterns of family inheritance, but the work gave the family greater stability. Like women elsewhere in

America and England, Chester County women displayed more testamentary freedom than did men because they had different responsibilities and therefore could respond to the needs of near-kin, whereas men took care of nuclear kin by leaving them almost their entire estates. Women's bequests cemented the broader kin base so necessary in times of change.[33] Thus women played an integral part in maintaining the patriarchal family in the eighteenth century.

CHAPTER 3

Farm Household Labor

It was early July, mid-way in the annual farm cycle, the beginning of the harvest. The rye, in gray-green sheets, stretched across the rolling hills. Farther on, the thick yellow hay waited for the scythes. Flax stood ready for pulling. Oat fields were still green, the buckwheat just covered the fields, and broad deep green leaves of corn were knee-high. The potato plants were just blossoming. Along the fields, split-rail fences marked the boundaries set against wandering cattle. Beyond these fences worming through the countryside, narrow roads cut through woods to log houses, a few barns, and here and there more substantial stone houses. In the farm kitchen yards, the women were stirring, bringing pails of warm milk from the morning milking, setting out crocks of butter from yesterday's churning, fresh loaves of bread, cheese, and jugs of rum.[1]

At sunrise, young neighbor women began to congregate in a kitchen yard, joining the men in preparation for the rye harvest. The morning "piece" eaten, they paired off and moved out to the fields. A dozen or two workers might congregate at the larger farms, but on most places two or four hired workers composed the harvest work force. The women were dressed in ankle-length light linen petticoats with separate tops called short gowns, with kerchiefs over their heads. They carried wooded-handled sickles with long curved eighteen-inch blades. One side of the blade was cut like a file, biased very fine, the other ground smooth so that the sharpened edge was serrated and could cut grain without slipping.

The rye had been planted in six-foot "lands," and the women and their male partners now began to move down the lands. Each team cut half a land, each reaper half that much. Grabbing a "grip" of the waist-high rye, the reapers took a cut first to the right and then to the left, holding the grain in their left hands and putting it behind them in a single pile. The next pair took the next land to the right, keeping about eighteen inches behind, thus moving through the field in rhythmic waves. At the end of the row the reapers shouldered their sickles, moved back down the rows, gathered the grips into sheaves, and bound them.

Haying followed as soon as the rye was gathered into house or barn. Here the men swung the heaviest scythes and women followed behind, spreading the grass to dry. Later they worked with the men, forking it into ricks or onto wagons to carry it back to barns.[2]

Then the women worked on, usually alone, pulling flax in the fields. They first took off the seed and then spread the flax in the field or a pond to rot the stalks (called "retting") so that the fiber could be separated from them. Later the men helped dry the flax over an open fire outdoors and break, or "scutch" it. Next came "hackling" to separate the broken stems from the fiber, a process sometimes put off until winter. In winter also, women separated shorter tow from the longer fine fibers and combed them into bundles, or "stricks" ready for spinning. Flax processing thus was not finished until January. Flax pulling was often done cooperatively, with the woman whose field had been pulled providing an eveningof food and entertainment as pay. The other cutting was usually hired, sometimes with farm daughters joining the hired hands in the fields.[3]

Other crops were harvested in fall in addition to grains and flax. Potatoes had to be dug, hops and apples picked. Cider making immediately followed apple picking; so too did apple drying. In November, hogs were butchered, the men normally doing the killing and cutting and women preserving the meat in brine, and rendering the fat for tallow and lard. Some women made candles and soap now, but less affluent households still used a rag in grease for winter light and did not use soap. There was little meat on the tables of these families and little fat on their hogs.[4]

Winter was the time to complete the flax processing, to spin it and card the wool, to weave or ready the yarn for the weaver, to bleach the woven linen, to dye cloth with red maple or sassafras bark, to sew simple clothes or hire a neighbor woman to help. Linen could be spun sitting down but wool had to be spun at a walking wheel, a process that involved repeatedly walking fifteen to twenty feet to draw out a rolag. The manufacture of clothing was an incredibly time-consuming process. As often as possible old clothing was patched and darned. Time was taken, too, to teach young children to read; the Bible and the yearly almanac were favored reading materials.[5]

Spring brought the planting of the garden, the freshening of cows, and the daily chores of milking, of making cheese and butter, of gathering plants for dyeing. Then followed the gathering of summer garden vegetables, pickling of cabbages and cucumbers. Nested within this yearly farm cycle were the cycles of monthly food processing and clothing maintenance. In one Birmingham household, women baked once a week and washed every two weeks. Some women ironed; most probably did not. Meals were simple, using foods processed earlier. There was little cleaning, for furniture remained plain and sparse. This was the farm cycle of women's work in the Brandywine Valley by the second half of the eighteenth century.[6]

The life cycle of women determined the extent of involvement in the full farm work cycle. Until the age of five, daughters had few work obligations. They mostly watched their mothers, allowed to imitate them but not expected to contribute to the work process. Yet by watching, they were ready to begin to learn a great many duties between five and ten years of age. During this period little girls learned to churn, to spin, to milk, to sew, and to read. After ten, at least by twelve, they assumed their places next to their mothers as full apprentices in housewifery. Some daughters might be working for others by twelve or fourteen. Young women went into households in winter to spin, do plain sewing, sometimes to weave, or to nurse or doctor on a jobbing basis. Spinning girls were most common in farm households, coming like shoemakers and tailors to spend two weeks working and visiting, bringing news of neighbors and community. In summer, these spinning girls performed extra harvest jobs. They hired themselves out for room and board and wages to help reap, do dairying, and assist with the heavy summer chores.

Economic status also affected the work cycle of women. A householder, one who lived on land owned by her family, usually arranged and supervised her own work subject only to the demands of her husband and other members of the family. This status as householder brought women considerable independence and visibility in historical records as well. Tenant women also arranged their own work, but they often sold part of their labor or the products of their labor to large householders. Women in artisan families sold part of their labor, too, if it was not needed at home. During harvest time, both tenant and artisan women provided an important source of hired labor. This work was often accounted for in record books, usually with simply a name, the type of work performed, and the amount paid. Considerably less control over their work was possible for "inmate" women, whose husbands contracted family labor in return for a place to stay on the farm of a householder. Contracts varied depending on how good a bargain could be struck in March, when these negotiations usually took place. Often, an inmate woman's work was an integral part of the bargain. She was explicitly obligated to help at harvest, with the dairying, or with animals, and sometimes to provide other household services as well. The labor of these women is least visible, for few of these contracts survive.

The wealthiest families and widowed males also had year-round female servants, some of whom in the second half of the eighteenth century were still indentured and others black and enslaved.[7] Indentured servants had provided some of the seventeenth-century colonial labor force. A system adapted to New World conditions from English agricultural servant labor, indentures allowed relatively poor rural men and women to use their labor to pay for passage to the colonies. The system transformed a one-year family servant into a long-term worker whose labor became a type of property to be sold or traded at will by his or her owner.[8]

The number of indentured women servants who came from England to the colonies was never large. Even in the late seventeenth century when planters seemed eager to purchase female servants and to marry them, the proportion was less than a quarter. By mid-eighteenth century their numbers had dropped to less than 10 percent. In a sample of servants leaving for the colonies in the 1770s, for example, several hundred left England for Maryland, but only fifty-four left for Pennsylvania. Most were young and unmarried.[9]

Agents did not encourage the importation of women in the eighteenth century. Some described them as "troublesome," although they never explained exactly why. Apparently a number of women became sick on one voyage. Pregnancy or inability to provide suitable references for household work might also have been elements discouraging the importation of women, according to Sharon Salinger who studied late eighteenth-century Philadelphia indentured servants.[10]

Women who came as indentured servants usually chose to work in the city. Almost 75 percent of the runaways in the late eighteenth century came from rural areas, although only half lived there. When servants had a choice, they chose urban work because it was easier, because it provided greater interest, or because the opportunities to prosper in rural areas were fewer than in urban areas. Whatever the cause, fewer came because conditions were not favorable. Apparently few older redemptioners were indentured in the valley. Jane Lilly, an Irish redemptioner, spent the four years from 1792 to 1796 paying off her passage from Ireland by working on a Willistown farm. She seems to have been an exception.[11]

The scarcity of indentured servants and the need for year-round help may have increased the demand for female slaves in the late eighteenth century. Distaste for slavery had not deterred Quakers from owning blacks in the early part of the century when there was an insufficient supply of workers. Alan Tully has estimated that at least 70 percent of the fifty-eight Chester County plantation owners who held 104 slaves in bondage from 1729 to 1758 were members of Quaker meetings. While some Chester County Quakers registered complaints with the Philadelphia Meeting for lack of action on manumission of slaves, others were expanding the institution into the peaceable kingdom. One early Wilmington Quaker remembered seeing a gang of twenty or thirty black people being driven past his door and his Quaker mother buying a boy of eight. Other Quaker women complacently willed these people as property to be inherited. When Mary Harlan died in 1741 and ordered her estate divided among her children, she still held Mingo in slavery. Betty Hollingsworth inherited "Him" from her husband Enoch in 1752. Tully feels that the Quakers held black people in bondage as a form of conspicuous consumption, because so many were women and children and worked in households of the wealthy. More likely, they were engaged in year-round domestic production.[12]

Yet women also shared in the manumission of slaves in Chester County. In the 1750s, approximately 9 percent of the males and 2 percent of the females leaving wills held blacks in bondage. Of a sample of males, all passed their bondswomen and men on to their children. One woman also passed on one-and-a-half-year-old Sarah to her son. A second woman at her death in 1750 manumitted Bella (black women were usually referred to only by their first name in wills) and bequeathed her a large number of household tools and furniture.

Bella's fortune may have been unusual, not because Quaker Deborah Nayle freed her, but because of Deborah's provision for Bellas's life as a freed woman. Bella received twenty-five pounds, a feather bed, Deborah's "best coverlids," three pairs of sheets and two pillowcases, her "best pine Chest with a Lock & key to it," earthen, wooden, and tinware, an armed chair, kettles, skillets, pots and pot hooks, two chairs, a couch, all her spinning wheels, flax and yarn, napkins and tablecloths, two books, baskets, boxes, all her wearing apparel except one gown that went to a friend, her best saddle and bridle, and all her provisions including malt, butter, and cheese. The bequest must have made Bella one of the wealthiest black settlers in Chester County, certainly in Thornbury Township.

Chester County Quakers took the lead in the early eighteenth century in trying to persuade church members to abstain from buying slaves. No official action by the Philadelphia Yearly Meeting followed, however, until the 1750s when the central body approved publication of antislavery tracts and began to urge Quaker meetings to take action against slave-trading members. Importation of slaves into Philadelphia peaked soon after, from 1759 to 1765. In fact, the use of slaves began to rise in rural areas as production for the market increased and female labor became scarce. But the Philadelphia slave market was a precarious business, plagued by a seasonal and limited demand. It proved difficult to maintain a black labor force, half slave, half free. Free urban blacks sheltered rural runaways. Black families, separated by sales, continually disrupted work by attempting reunions. The trade did, however, provide the basis for a considerable black labor force in the valley and for the establishment of black communities there.[14]

In Quaker communities, moral concern about slavery gradually prevailed. According to Jean Soderlund, who has made an extensive study of Quaker slaveholding, once the wealthiest farmers opposed slavery, it ended in Quaker areas. By the early 1760s, Quakers in both Chester and New Castle counties were active in freeing slaves, and by the end of the decade few slaves were left in Quaker households. Many black women were freed in the county during this time and were compensated for their work. Dinah, for example, a black woman enslaved from eleven to eighteen years of age in East Nottingham, received a house, bedding, free firewood, and half acre of land rent free for eight years when she and her two daughters were freed. By 1775, Quakers who refused to free their slaves could be disowned; thereafter, only Anglicans, Swedish Lutherans, and disowned Quakers held on to a few remaining slaves. In 1780, Pennsylvania passed its first act for gradual abolition of slavery. Blacks already in slavery could be held for life, but their children could be held only as indentured servants until the age of twenty-eight. The importation of additional slaves was prohibited. Thereafter, the practice gradually disappeared from the Brandywine Valley. In 1780, only five slaves remained in Kennett Township, and none remained in East Bradford Township. According to the slave registry of that year, the Brandywine Valley was almost free of slavery while townships bordering it still had from ten to twenty slaves each. By 1790, the black population formed a small free rural community.[15]

As black women worked their way out of bondage in the valley, other black women became more deeply mired in the unfree labor system to the south. Household inventories clearly show this shift. The number of enslaved black women listed in Chester County inventories declined from 9 to 2 percent between the 1750s and 1790s. During the same period in New Castle County, the number increased from 14 to 41 percent in male-headed households. Black women's labor south of the valley was thus essentially different from their work to the north. In the south most would labor as bondswomen, and in the north, as servants and day laborers (see Appendix tables 9 and 10).

Some black women—freed slaves, the children of slaves, and orphans—remained indentured. Just how many is difficult to determine. One family of four black girls—Jenny, Rose, Cynthia, and Cherry Morris—were indentured out of Philadelphia onto Chester County farms. A group of slave ships captured in 1800 were the source of some other indentured blacks. Of the 126 Africans who survived the voyage and the admiralty trial of the ships' owners, nineteen males and two females were indentured to Chester County farms. Mary and Demau, each fourteen years of age, served four years each.[16]

The indenturing of orphan girls by the overseers of the poor continued into the nineteenth century. The 129 Chester County orphan daughters (black and white) who were indentured between 1803 and 1825 were very young, usually between three and seven, and the 210 New Castle County girls indentured between 1827 and 1835 were between six and fourteen. All were indentured only until eighteen, however, so they provided only a few years of mature labor before becoming part of a free rural labor force. This group of workers took form when indentured servitude of adults disappeared at the end of the eighteen century. The women sold their labor to the farmers of Chester County on a yearly basis, by the season, or by the task.

In the Brandywine Valley, the decline of both female indentured and slave labor was hastened by the increase in the number of young native-born white women available to perform work on farms. Problems exist in accounting for their increase. Some householders listed on tax lists apparently were actually renters or sharecroppers, not landowners. Lucy Simler has argued that most accounts do not accurately describe the position of these people because scholars have considered the lists of heads of households to be proxies for land ownership rather than simply lists of taxpayers—both owners and those renting for cash and on shares. Lucy Simler estimates that over time 30 to 40 percent and in some townships as many as 50 percent of those assessed rented rather than owned land. Moreover, tax lists contain the names of male "inmates" whose family households included women who performed work for landowners as well as in their own households. And finally, many landless working people probably do not appear on any lists because they were too poor to pay taxes. Until new ways of estimating the population are found, descriptions of the work force must remain approximate.[17]

The problem of describing tenants, inmates, and the landless is difficult; yet these women may have made up an important part of the population. They were probably a very mobile group, moving from farm to farm and township to township, never settling permanently because they owned no land. Alice Clark was the daughter of one such family whose migrations the trustees of the poor recorded after she gave birth to two illegitimate children for whom the family could not care.

The trustees examined Clark's father, Joseph, to decide which township was responsible for supporting her and her children. This is the story Joseph Clark told. He was born in East Bradford and worked on farms in several townships while still a minor. When he reached twenty-one, he hired out by the year for four or five years on different farms. After his marriage, he worked for shares for a few years, paid rent for seven years, and then returned to working on shares. In most places, whether sharecropper or renter, he paid his poor tax, although one year, he admitted, he didn't "remember to pay Poor tax said year." Joseph was, at the time of his examination in 1785, paying rent and the poor tax, but he had no lease. He signed with his mark. These were the families who were often on the roads on April 1, searching for better terms as tenants. John Sugar of West Bradford remarked in 1793 that April 1 of that year was "remarkably favorable for Flitting families" and that thirty wagons were said to have passed through West Chester. Alice Clark and her mother would have been part of such a "flitting family," working on successive farms as their sharecropper-renter family moved.[18]

The work of the inmate women is just as difficult to determine. Each tax list contained the names of male inmates who paid no property tax but rather a tax on personal property. Most were married heads of families and they often had cows listed, a sign that a family, not just an individual male, was being enumerated. A 1767 list of five townships—East Bradford, West Bradford, New Garden, Birmingham, and Kennett—showed 113 inmate households, almost all of whom paid a property tax. In 1783, when the numbers of people in inmate households were listed, their households were, on the average, smaller than those of property taxpayers. In these same five townships, the size of 58 percent of all inmate households was reported and had an average size of 4.7 persons, whereas the figures for propertied households was 5.6. Daughters and wives of inmates provided part of the work force for their landlords as well as working at home. Since many of the inmates were also artisans, women must have had a major role in assisting their husbands with craft work as well as with keeping cows and gardens. Over half of the inmates kept cows in most townships in the 1780s. The Chester County Agricultural communities already had well-developed artisan and merchant classes. New Garden Townships, for example, by 1769 had twenty-seven artisans along with seventeen mill owners, three tavern keepers, and a doctor. Forbes estimates that about a third of New Garden Township residents were landless in 1774.[19]

Some landless workers often do not show up on any tax lists because their income was too small to be taxed. These are the most difficult households to describe because they appear in accounts merely as working for others. They do sometimes appear in artisans accounts, however. John Buffington, Jr., a West Bradford shoemaker, kept careful accounts in 1772–1773. But only 17 percent of the surnames appearing in his accounts also appear on the tax lists for those years.[20]

One must conclude that a large number of women performed a variety of tasks in payment for the use of land, for commodities, or for cash wages. Whether from tenant, inmate, or simply untaxed landless families, these women provided an essential work force for the developing agricultural economy of the late eighteenth century. The extent and variety of their work can be described in part through diaries, farm records, and traveler's accounts.

The most detailed of these accounts is found in the diary of Benjamin Hawley. Four months after his wife died in November 1761, he began teaching school at his home. Hawley was not born a Friend, but he apparently attended meeting and soon joined the Society. His six children were scattered along the banks of the Brandywine in Birmingham and East Bradford townships.

Among the women whose work Hawley described from 1761 to 1782 were several members of the Dunkin family.[21] In mid-March 1762, Hannah Dunkin went to work for Hawley on his Birmingham farm for three shillings, six pence a week. It is not clear where Hannah lived, for tax accounts from nearby townships do not contain any Dunkins. She may have come some distance, or she might have belonged to an unlisted family. At any rate, Hannah was from a large family, and when she wanted time off, another female Dunkin usually stayed in her place. Ann Dunkin, perhaps her mother, sold Hawley butter. Hannah worked until December of 1762 and then went off to work for a week for neighbor George Martin, her sister Sarah replacing her. After her return on December 27, Hannah was put to bed with some unknown sickness. Rachel Dunkin came to nurse her and left when she seemed better. Hannah washed on January 9, and two days later she was back in bed. Rachel again came to care for her, gave her a physic, and returned home.

When Hannah still did not recover, Rachel came yet again and began staying overnight. Hawley himself went to another neighbor for physic and finally supplied Hannah's father with money to send for a doctor. Rachel was now at Hannah's side almost every day, nursing and taking over the household tasks. Hawley sent for Ann Dunkin to come. Hannah seemed near death on March 12, and her father, mother, and several others came to visit. Six days later, Hannah asked that one of the Quaker ministers come home to have a meeting with her, and neighbors came in to visit. Finally, on March 21, neighbor men carried Hannah home on their shoulders. Hawley called to see Hannah two days later. On March 29, Rachel went home and Sarah came to work for Hawley. The next day Hannah died. She was buried on April 1 at Birmingham Meeting with Hawley present. Sarah worked at Hawley's farm until August 31, when she too left. Hawley continued to buy butter from Ann Dunkin for the next six years. Rachel was back working for Hawley in 1771 for two weeks and Sarah spooled flax for him in December.

The relationship between the Dunkin women and Hawley over almost a decade closely reflected the nature of the work that joined neighbors in economic relationships in rural Chester County. The Dunkin women performed essential types of work: they boarded to perform year-round work, took in textile work, and sold processed food. While daughters worked out, their mothers worked at home processing food and textiles. Both mother and daughters were part of a community willing and able to work for their wealthier neighbors. We really do not know if there was a stark line between employer and employee, but from this account they seem to have been part of an ebb and flow of economic and social life, tied by both work relationships and other relationships that existed. Class differences and conflict may have also been an accepted and integral part of this community, merely muffled by the dependence that individuals had on one another.

Hawley's account also listed work by neighboring women other than the Dunkins. He noted on June 5, 1771: "Rachel Seal came here paid her 2/6 & 1/2 a Quire of paper 1s for making 3 caps & 2 hoods & 6d for Bleeding Betsy make total in Cash 3s." This was the same Rachel Seal mentioned in chapter 2 who practiced medicine in Birmingham. On January 21, 1771, Hawley reported that Nancy had been there twenty weeks at 3/6 and that after she left Betsy arrived. Later he wrote of paying Mary Mullen to wash, but after nineteen weeks, she became unwell and Mary Lion came to bake and wash; he paid her six shillings. Mary's husband also worked haying, and after twenty-eight weeks of employment, Hawley paid off both Mary and her husband. Patience Lion then came to work.

Richard Barnard, a farmer nearby in Bradford Township, similarly reported hiring women on his farm in the 1770s. Mary Jones worked from February to May 1775. Mary Crofton came to nurse for seventeen days at seventeen shillings. In July 1776, Barnard reported "pooling flax my hands Elie & Jane," and later he paid Jane Porter for reaping wheat. Other women worked for Barnard in their own homes, mostly processing textile fibers. Mary Bakes dressed eleven pounds of flax for one pound, eighteen shillings. Philis Eckhoff bought cider and a bushel of wheat and paid for it in spinning. Hannah Harlan received six pounds from Barnard to teach his daughter to be a "tailoress." Hawley also talked of paying for similar services, leaving thread to be whitened at the house of one neighbor, paying Hannah Woodward for weaving inkle, and engaging Sarah Dilworth to make a gown for his wife before she died.[22]

An account kept by William Smedley of Middletown from 1751 to his death in 1766 and then by his wife Elizabeth also listed a great variety of work done by women for pay. Most frequently women spun, but they also reaped, sowed and pulled flax, and gathered apples. They washed, made shirts and trousers, and took in boarders. They sold butter, cheese, eggs, candles, stocking yarn, soap, and featherbeds. The jobs mentioned are frequent and varied enough to indicate that there was a strong demand for women's work in the late eighteenth century on Chester County farms and that farms were highly interdependent.[23]

Descriptions of the farm work cycle and accounts of women performing different types of labor are very informative. They provide enough evidence, for example, to call for revision of some historians' conclusion that women did not work in the fields. In addition to the above accounts, other exist. Travelers, who as outsiders did not give detailed accounts, nevertheless made occasional references to women's fieldwork. John Watson noted "many men and some women" sickling wheat and rye in the 1740s, and in 1749 Joshua Henstead reported Irish women reaping in New Castle County. Benjamin Rush noted in 1789 that "*Baurmaedchen and Baurbursch*" worked together in the fields. A 1771 account listed the wife of a laborer as receiving payment for reaping. An account from 1795 mentions women being paid for pulling flax. Other late eighteenth-century writers reported young women hoeing and turning hay and early nineteenth-century accounts continue to show a great variety of tasks performed by women including fieldwork. An account book kept by Martha Lewis between 1827 and 1835 shows women paying for provisions by spinning, sewing, knitting, cleaning house, whitewashing, cooking, butchering, pulling and hackling flax, planting and cutting corn, and making hay. In addition some women simply worked by the week at unspecified tasks.[24]

Although more accounts will undoubtedly come to light, some women's work will never show up in the records. Inmate women—both wives and daughters—were expected to work in the harvest, so that the work would not have been recorded. They were not paid for it separately, for the inmate's contract with the landowner included all the family's labor.

Thus, the accounts, although they may eventually provide more numerous records of women being paid for harvest work, will never fully explain how women's fieldwork varied in amount over time. One possible approach to this question is to study changes in agriculture technology. The sickle and scythe were the main reaping implements used through most of the eighteenth century in western Europe and the American colonies. The success of harvests rested upon the workers ability to use them skillfully. In the eighteenth century, the appearance of a new tool began to change harvest labor patterns dramatically. It was a device called the cradle, a rakelike appendage that literally cradled the grain as it was cut by a scythe. Its use spread through Europe after the first quarter of the eighteenth century. Within twenty-five years, some areas had shifted to the cradle in 50 percent of reaping and mowing. Within a hundred years, the sickle had virtually disappeared as a harvest tool except on farms with difficult field conditions, those with stones, tree roots, hollows, or where labor costs remained low. As agricultural markets developed, farmers steadily switched from sickle to scythe and cradle.[25]

This transformation of harvest technology is important for understanding women's farm work, because the scythe was a gender-specific tool, whereas both men and women used the short sickle equally well.

To harvest with it, the reaper assumed a deeply stooped position, grasped the grain in one hand, and neatly sawed it. The scythe, with its long curved handle and thirty-inch blade, was heavier and required the worker to employ a quick glancing stroke with the sharp blade. Cradles added more weight to the scythe, particularly in the American colonies where they were larger than those in Europe. The long scythe, then, with its heavy cradle attachment, was more difficult for women and for short men to manipulate.[26]

Physical ability may not have been the only factor in the scythe becoming a gender-specific harvest tool. Scythes, an ancient tool, came into common use in eighteenth-century England as farmers began to furnish them to the youngest and strongest harvesters on large farms. The scythe was cost efficient, allowing much more rapid mowing. A reaper with a sickle could harvest a fourth to a third of an acre per day if binding and shocking were included, and about one acre if they weren't. With a scythe, a harvester could do over twice as much, depending on the type of grain. Grain used for fodder, such as oats, was the first to be harvested with a scythe because the condition of the grain was unimportant. Buckwheat and wheat could be harvested with a scythe and cradle. Rye continued for some time to be reaped with the sickle because more of it was lost if a cradle were used. In Chester County rye probably was second only to wheat in quantities harvested during the last quarter of the eighteenth century. But its disappearance as an important grain soon after the turn of the century coincided with the widespread use of the cradle. During periods of technological change, men may attempt to maintain certain types of technology as male specific when new tools reduce the number of workers needed and increase wages. A similar process may have occurred in the grain harvest.[27]

The shift from female to male harvest labor and from sickle to scythe is clearly documented in England. Pastoral areas switched first because large areas needed to be mowed. Women shifted their work cycle from harvest to spring, as they moved into dairying. In areas where grain remained the principal crop, women had more difficulty finding alternative farm labor. English women often migrated to urban areas or moved into textile processing.[28]

Although less research has been done on harvest technology in the American colonies, a similar transition from sickle to scythe took place. No clear pattern of the switch is yet evident. Some sort of a scythe with a cradle was introduced into the South in the mid-eighteenth century. It may have been adopted there first because stands of wheat were lighter and thrashing was done in the fields. There were also better alternative uses for women's labor—both black and white. Textile tools increased in number in the South about the same time that cradles appeared. Black women continued to work as sicklers in the field long after males were at work with scythes and cradles, however. Thomas Jefferson had both males and female sicklers at work on his plantation along with male cradlers in 1795.[25]

The use of scythes spread quickly in the late eighteenth century. In the 1750s, cradlers or scythes and hangings show up in less than 5 percent of the New Castle and Chester County inventories. But by the 1790s, they appear in inventories for over 40 percent of farms in Chester County and over 20 percent in New Castle County. Cradles did not completely displace women in the fields for another two decades in Chester County. Nonetheless, the transition, moving from South to North, may have accelerated during the first decade of the nineteenth century. As women gradually moved into other occupations, their labor was less available at harvest time and the scythe was adopted. In the Mid-Atlantic region, women may have moved out of the fields into textile processing and then into dairying as farm families turned from grain growing.[30]

If women worked less in fieldwork during the last quarter of the eighteenth century, they found another type of outdoor work had increased—work with animals, which like other types of farm work, was gender based. Women sometimes raised cattle, especially cows, but males were usually responsible for feeding and bedding them. Women, however, inherited two principal animal-related duties in the Mid-Atlantic—milking cows and raising poultry. (Women's work with cows will be discussed in later chapters that deal with butter making.)

Tracing poultry growing is difficult. Unlike cows, poultry was seldom valued enough to appear on the inventories of Chester County and New Castle County households. Poultry appears in only 3 percent of almost three hundred inventories sampled from the two counties in the 1750s and 1790s. Although travelers' and household accounts both refer to hens, ducks, turkeys, and geese being raised, there is no way to measure the sale of eggs and poultry before the 1770s, as few sales records exist for this period. There is indirect evidence, however, that market flocks increased dramatically in rural areas in the last half of the century—featherbeds became items of importance and value in the households of the colonials (see Appendix tables 9 and 10).[31]

Before the 1750s, households had few featherbeds. Women commonly used shucks or chaff to stuff mattresses, replacing the contents every year. Although they were often inventoried, these mattresses were not valuable. Jack Michel, for example, reports the simplicity of Chester County bedding in the early eighteenth century. By mid-century, however, featherbeds show up frequently in the New Castle County and Chester County household inventories. Fully a third of male-headed households in New Castle County and almost half of female-headed households in Chester County had them. By the 1790s, although chaff mattresses still had not disappeared from inventories, all types of households had large numbers of featherbeds. As attested by their weights in inventories, they were sometimes enormous. The Maxwell household in New Castle County had three featherbeds, weighing thirty-four, fifty-six and sixty pounds. And Mary Jenkins, a Chester County widow, had a whopping seventy-six-pound featherbed to lie on when she died. The value of the feathers made these beds worthy of being passed on from generation to generation. They were not only a common inheritance item; they appear frequently as a special bequest to a child or other relative. Husbands often left to their wives the featherbed the couple had occupied. Daughters and sons received best featherbeds in addition to money bequests. Bedsteads frequently accompanied the beds, but in most cases it was the feathers that gave them their value. Raising, butchering, and plucking fowl and then saving the feathers became a part of women's productive work.[32]

Feathers and featherbeds were only one small part of what Cary Carson and Lorena Walsh have called the consumer revolution of the eighteenth century. In at least three major ways women were involved in this revolution. They were primary producers of great amounts of textiles and clothing. They increased the amount of food available for home consumption throughout the year by devising elaborate methods of food preparation and preservation. And finally, they were responsible for the management of increasing amounts of material objects in addition to tools for food processing and textile manufacturing. As status began to depend on these material objects, the role of women as managers of the domestic space within which tools were used and objects displayed became more important. Rural colonial women devoted increasing amounts of time to domestic management.[33]

Although food-processing tools were important adjuncts to women's labor, these tools were quite simple in the 1750s. The hearth, still the center of processing tasks, held fire tongs, racks and hangings, pots, earthenware, and pewterware. Most households also had frying pans and dough troughs to prepare bread. These tools were common, but less than half the households had tubs, pails, and churns listed in inventories and less than a fourth had tinware, cheese presses, or baskets (see Appendix table 9).

Wealth normally determined the amount of textiles produced in households. Only 10 percent of the households, probably the wealthiest, had looms. By far the most common textile tool was the linen wheel; wool wheels were much less common. Thus, producing linen yarn must have been the most common task for women. Most of this yarn was sent out for weaving and then returned to the household for bleaching before being turned into linen for the family's use. By the 1750s, the amount of linen needed to run a wealthy household was already large. There were bed ticks to make, at least one for every two people. About a third of the households possessed table linen and almost a fourth had sheets.

The wealthiest households used pillowcases and bolster covers, coverlids, and curtains for bedsteads. Few households in this period had quilts because both linen and wool were expensive and time consuming to make, and there were as yet no cotton cloth scraps and filling that would make quilts a favorite nineteenth-century textile. When women headed households, they usually increased the amount of finished textiles in their homes. Far more women than men had featherbeds, sheets, table linen, bolsters, pillows, cases, blankets, coverlids, and curtains; some added quilts and rugs. These stockpiles of household textiles grew impressively. When Mary Roberts of Radnor died in 1750, for example, although she had only one featherbed, she owned fifteen sheets. Most households kept on hand a supply of yardage, usually linen, and women frequently produced linen yarn into old age.

Because textile processing was at the center of women's work, it is not surprising that fiber arts became their special artistic creation. By the end of the eighteenth century fancy sewing had developed into a fine art, practiced and perfected by women who had the leisure to engage in more than plain sewing of sheets, tablecloths, and napkins. Although Quakers generally discouraged fancy sewing, surviving artwork from Chester County shows that young women embroidered samplers in schools, completed richly decorative crewel work, and quilted single-color unpieced coverlids.[34]

Most women did not yet have large amounts of furniture that needed much attention—what might be termed service rather than production. Large numbers of households inventoried chests, books, tables, and chairs. Most households had a Bible, and many Quaker families also had specialized Quaker tracts. As for chairs and tables, some households accumulated an amazing number of them. Sarah Robinson, who died in Chester in 1754, for example, had twenty-nine chairs and eighteen tables of various sorts.

Aside from the care of tables and chairs, however, the service demands of even the wealthiest rural homes were minimal. Cleaning looking glasses was perhaps the most common, for almost half of female-headed and almost 40 percent of male-headed households had them. The mirror, as Michel Foucault reminds us, is a device for replicating the contents of a space, but also for allowing the people within a scene to see it as if they were outside. Although in the seventeenth century looking glasses were mostly a luxury for the nobility, by the next century Philadelphia had looking glass stores and mirrors adorned the corners of rural households. They were always called looking glasses, not mirrors, by the inventory takers of the Brandywine Valley. There was spectacle here, in houses with no pictures and few window curtains or rugs.[35]

If one could peer into the looking glass of one of the 40 percent of households that had them, one might have seen the household of Hannah

Oskidlen of Pikeland in 1755. The Oskidlens owned a three-room house, with an upstairs containing three beds and a back room on the first floor holding a trunk, a chest, another bed, a desk, and some small items. Crowded into the front room was another bed, bedding, and dresser together with all Hannah's equipment for processing. At the hearth were fire tongs and shovel, hooks, pot hangers, three pots, a skillet, a gridiron, and a dough trough. The room held her churn, pails, a tub, and two meat barrels, as well as a large wool wheel and small linen wheel with a check reel, riddles, spools, and wool and flax. The inventory noted sixty-seven yards of linen at the weaver. Amid her domestic tools were five chairs and "a little one," a lantern, and a looking glass. And in the looking glass, one could have seen Hannah, "now far Advanced in her pregnancy"; Hannah who like her husband Morris could not read or write; Hannah who, amidst the clutter of her small world, nursed her husband in his last illness with the help of a servant girl, still indentured for two years and eight months; Hannah to whom Morris would leave a third of his estate and who, within three years, would be remarried.[36]

Accounts of the Revolution and depredation claims provide another way of measuring women's work. Two armies rode into the peaceful valley in September 1777. General Sir William Howe landed at the mouth of the Elk River with an army of eighteen thousand and entered Chester County on September 8. Three days later, near Kennett Square, Howe divided his army, sending one group up the old Chester Road to Chadds Ford. At the head of the other group, he crossed the western and eastern branches of the Brandywine and outflanked the American army assembled at Chadds Ford. After hard fighting around the Birmingham Quaker meetinghouse, the Americans retreated toward Philadelphia. They attempted a skirmish on September 16, but after a hard rain the armies separated again. Washington sent General Anthony Wayne to harass the British while they recrossed the river. Completely surprised by the British, Wayne's mission failed, thus opening the way for the British to advance on Philadelphia while Washington went into camp for the winter at Valley Forge.[37]

Howe's purpose in cutting through Chester County was to sever North and South, capture American stores of munitions, and pacify the countryside. Instead he severed Philadelphia from its hinterland, leaving the rebels battered but still in control of the countryside and their source of supplies. Had either side sought to engage in guerrilla warfare, farmers would have suffered severely. As it was, the British first appropriated what they needed from the prosperous farms; then the Americans in turn taxed and requisitioned, usually taking farm animals and produce in payment. It was, according to Quaker chronicles, a time of suffering.

Some farmers supported the war, but Quakers in the rural backcountry could not support the war without giving up their faith. Thus many remained stubbornly neutral in a war in which neutrality was seen as treason. Farmers were particularly vulnerable in an era when armies lived off the land. Their visible wealth made them targets first of the British as the army foraged for food and portable items and then of the Americans who taxed the Quakers for not sending their sons to war, often taking the fattest calf in payment. The succeeding accounts of war taxes and depredation claims give us a picture of rural Chester farms we might not have otherwise.[38]

Farm families had prospered from a long upturn in the economy from 1745 to 1760. Exports from Philadelphia had grown remarkably during those years—some items by over 120 percent. A population boom had also increased demand for foodstuffs at home. Although economic development generally slowed after 1760, the French and Indian War that ended in 1763 had further extended economic opportunity for the rural population; the war economy had quickened their involvement in commercial agriculture and brought new consumer items to isolated and relatively self-sufficient households. For example, tea was not common in the backcountry before the 1760s; now countrywomen experimented with the dried leaves or asked more affluent neighbors for advice. Does one boil, butter, and eat? *Steep* entered the rural vocabulary along with increasing numbers in their homes of teakettles, teapots, tea canisters, teaspoons, teacups, and tea tables. The spread of this tea-making paraphernalia ("tea equipage" it was sometimes called) was truly astounding. More important, farm families simply accumulated more of everything. They built more substantial stone houses and increased the number of their animals, feeding them better by irrigating meadows and supplying fodder year-round. Because of the continued growth of domestic demand, the export of wheat, and the increase in textile production, the 1760s and early 1770s, before the coming of war, were still good times, relatively speaking.[39]

Claims made by fifty-eight farm households after the Revolution, most from around the Birmingham area, reveal some patterns of women's production not revealed by accounts and inventories. The inventories, for example, almost never mentioned fowl, but on depredation claims, seven flocks of twenty to almost one hundred fowl appear. These are numbers large enough to indicate market production of poultry and eggs.[40]

Another industry that claims indicate had become an important source of income was dairying. Eleven households reported losing one hundred to two thousand pounds of cheese, and two Brinton households in Birmingham listed three thousand pounds. The presence of earthen milk pans, pails, and tubs indicate the existence of a cottage industry. There was relatively little surplus cloth or butter, although a number of churns was destroyed.[41]

Besides the utensils and products of cheese making, women lost considerable amounts of clothing. Everything from bonnets and stays to pockets and needles disappeared during the battle. Women listed gowns, short gowns, petticoats, shifts, cloaks, stockings, shoes, aprons, caps, and handkerchiefs among missing apparel. Three women in Elizabeth Davis's household in Haverford lost clothes, not in battle, but later in the year when British troops visited their home and helped themselves to everything from sheep, hogs, oats, and wheat to three pairs of her scissors, spectacles, coverlids, blankets, sheets, and other farm items amounting to over 144 pounds in value.[42]

For these women and their families, the invading army made a lasting impression. In addition to cheese, churns, and butter-making equipment, the source of milk also disappeared. Farms lost over one hundred cows in the invasion, a great loss since this was one of women's primary concerns and a source of productive work. It is not surprising, then, that one of the stories handed down through generations of Quakers concerned an encounter between Jane Gibbons of Westtown and General Howe. She asked him to return her favorite cow, and he replied, "I'm afraid, madam, that you love your cow better than your king." Such folklore reinforces the image of Quaker women standing up to the British and maintaining their neutrality.[43]

But no stories have come down reminding descendants of what Quaker women said to Continentals when a few years later the new government fined Quakers for the expense of the war and for not appearing at militia muster. Quaker women certainly could not have been happy about parting with forty-five cows and twenty-three heifers along with sheep, hogs, wheat and other farm produce.[44]

The tasks performed by farm women made them absolutely essential to the success of any agricultural venture. Without a family, a man was likely to have a difficult time, as Brissot de Warville sadly noted in 1788 after visiting the French bachelor M. Le Gaux who farmed on the Schuylkill River thirteen miles north of Philadelphia. "As he is without a family he does not have any poultry or pigeons and makes no cheese," Brissot de Warville wrote, "nor does he have any spinning done or collect goose feathers. It is a great disadvantage for him not to be able to profit from these domestic farm industries, which can be carried on well only by women." And by women of the family, he might have added. For Le Gaux employed a German indentured woman, but even with an eight-year-old girl to help her and the added incentive of a share of the proceeds from the sale of calves, she was not able to handle the work successfully.

On the other hand, a visit to a Quaker farm near Middletown, twenty-two miles from Philadelphia, pleased Brissot de Warville immensely. Here the women not only provided him with "a good bed, snow-white sheets, and a fine counterpane"; they also gave evidence of their "family industries" through the production of vegetables, butter, and cheese sold in town once a week and the spinning of wool and linen. The mother and five daughters had enough work to occupy them constantly while three sons worked in the fields. "It was truly a patriarchal family," Brissot de Warville wrote with enthusiasm.[45]

In the 1790s, much of women's work remained relatively unchanged. In textile processing, women still worked primarily at their small linen wheels. But the cotton that appeared in some of the inventories was a small predictor of the revolution the new fiber and its processing would bring. The amount of textiles women produced as a cottage industry peaked during the late eighteenth century. But cotton was more difficult to spin than flax or wool, and women would soon be spending their time at more productive tasks (see Appendix table 10).

Food processing had not yet reached its limits as a cottage industry, and the amount and variety of tools increased in the 1790s. The basic tools did not change dramatically—hearth cooking tools, the dough trough, the earthenware, the pewterware remained. But there were far greater quantities of other tools now—more frying pans and kettles, griddles and gridirons, churns, pails and tubs, tinware. Most important, new tools—the oven and stove—appeared. By the 1790s, almost half the households had these essential energy-efficient tools.

The stove was a major household innovation. Although it could still cause burns, the hearth fire now was contained, lessening the danger of house fires. Six-plate and ten-plate stoves were common, which allowed a greater variety of equipment to be used. More equipment, of course, meant more scouring and more work servicing the tools to produce the food. But the switch to stoves may have had an even more profound significance. Women changed to them primarily because they required less wood for cooking. The open hearth had been a voracious consumer of wood. Le Gaux already realized in 1788 that he had to employ wood sparingly because of its growing scarcity. Although travelers in the 1790s noted the great variety of trees in the Brandywine Valley—red cedar, oak, hickory, black walnut, chestnut, tulip, poplar, ash, red maple, locust, elder, laurel, birch, willow, bayberry—even these forests could no longer provide the quantities of firewood needed for household tasks. Stoves may have been expensive but they were durable, and the alternative was buying ever more expensive wood from merchants like Morgan Reese of Chester County who owned over a thousand cords of wood purchased from seven farms. And as yet another harbinger of change, the Clendenin household in Oxford Township in 1797 recorded the first instance of an energy source that would soon replace wood in the once heavily wooded Brandywine Valley—coal.[46]

Perhaps the biggest change, however, was in the increased number of households possessing clocks. In the mid-eighteenth century, women's time had been task oriented. Life and work flowed together in what E. P. Thompson has called the "patterning of social time." The average rural Pennsylvania household had a relatively spare culture that conserved scarce resources. Labor patterns were irregular: intense labor alternated with idleness. Domestic servants participated in this pattern to a greater extent than hired labor, who traded tasks for money and worked more intensely—or so many farm householders tended to think in the late eighteenth century. Some scholars have argued that women retained pre-industrial work patterns longer than men, but women too quickened their work as commercialization increased. They too watched the clock—in fact, they were probably more conscious of it because they worked closer to it. By the 1750s, clock and watchmakers were advertising their skills in Philadelphia. Of eighty-five households sampled from Chester County during the same decade, 19 percent already owned timepieces, although in New Castle County, only one of forty-two households had a clock. But, by the 1790s, the use of clocks had spread southwards, and 35 percent of all households owned them. To city people, rural women seemed to work at a leisurely pace, but their work too began to increase as the tempo of the marketplace affected the household more directly. There is the possibility, of course, that purchase of clocks may not have followed from an increased need to know the time. Rather, they may have been a cultural necessity, a visible indicator of power in the community.[47]

In any event, farm women now watched the clock as well as the looking glass. They cooked on the stove rather than the hearth, and on middling farms, they presided over increasing amounts of equipment to make and serve food. A few were beginning to acquire chamber pots, candle stands, carpets, window papers, and other amenities.

Some women had more material goods and more money than others. Take the household of Charity Garrett, for example, who died in 1799. She left a comfortable estate of 500 pounds to two sons and three daughters. They divided notes, bonds, and a house crammed full of tools and goods. Into the three downstairs rooms of Garrett's house were stuffed not only all the processing tools of the time but also rocking chairs, baskets, stoneware, flat irons, religious books, a bureau and a tea table, over one hundred yards of cloth, a japanned server, twenty-five sheets, fourteen pillowcases, fourteen chairs, and six quilts. Charity Garrett was only one of a growing number of women who managed estates. Alice Peirsol left an estate of 1,655 pounds when she died in 1790 (see Appendix table 11).[48]

In this Philadelphia hinterland, as commercial agriculture was developing, classes were also taking form within which women used their labor in vastly different ways, depending on their economic status. Thus while tea equipage and the leisure to use and enjoy it was spreading among some women, other women were contracting out their labor more frequently. If wealthy farm widows were examples of independent women in Chester County, to the south there were increasing examples of unfree women living within wealthy male-headed households. Jane, Silva, Hetty, Hanna, Nancy, Rebekah—only their first names remain to remind us of their bondage in the households of wealthy New Castle men. And increasingly evident in both counties were women who did not fit into wealthy homes at all—as mistress, servant, or day laborer.

CHAPTER 4

The Social Geography of Dependency

Quakers in Chester County and New Castle County had always cared for their poor on the assumption that all are God's children, and we should care for one another. The men's and women's business meetings handled welfare informally at first. They collected and dispersed money to the poor, including widows, orphans, the aged, and ill, who had no relatives able to support them. Quakers built their own almshouse in Philadelphia in 1713 and expanded it in 1729, and rural Quakers continued the old informal system of welfare, soliciting money for the poor into the 1780s. As the Quakers became a smaller minority of the population and unattached women needing assistance increased in the late eighteenth century, Quakers began to lend their time and money to expand the rural public welfare system, one that would rest not on culture but on geography. Quaker males, like other males, moved into the public sphere to serve as overseers of the poor, becoming officials at the town level in Pennsylvania and Delaware. Poverty gave a woman the right to demand public assistance; but it also gave the public the right to control her and her children.

The geography of poverty brought poor women in touch with the township, the major source of public welfare in the Brandywine Valley. The township was the main community settlement pattern in Chester and New Castle counties. Rulings of the provincial government, together with the desires of the colonists, had interacted to determine the township form of community settlement. To encourage compact settlement William Penn had instructed that settlers take up five hundred acres each in township areas of approximately five thousand acres. These townships were laid out in a belt that circled west of Philadelphia and stretched back into the hinterland. By the end of the seventeenth century, eighteen townships formed Chester County. In the early eighteenth century, new townships were carved out of the old and added to the county until it had become an unwieldy administrative unit. In 1787, the eighteen eastern townships became Delaware County and the new western townships retained the old name of Chester County. In southeastern Pennsylvania, while the old unwieldy county was still in the process of division, the townships became increasingly important governmental units. The first rural public welfare institution developed at this level.[1]

Townships were not visible units punctuated by villages with town halls. Rather they were administrative units with constables, overseers of the poor, supervisors of highways, and movable meetings. A list of public officers for Kennett Township indicates that overseers were not always necessary in the early eighteenth century. While constables were selected in all but four of the seventeen years between 1705 and 1721, only six overseers were selected. Beginning in 1722, however, overseers of the poor were selected in all but one year, indicating an increased need to assist the poor, and after 1738, each office was filled every year. Overseers administered the Pennsylvania poor law of 1735, which with some amendments remained in force until 1771. Delaware, meanwhile, was separated administratively from Pennsylvania in 1701 but developed a geographical unit similar to the township called the hundred. In the three lower counties, the area that would become the state of Delaware, the county remained the primary local unit of government. But even there, the hundred had important local tasks, such as collecting taxes and overseeing the poor. The lower counties passed their first comprehensive poor law in 1731.[2]

The number of poor, particularly poor women, needing assistance increased in the 1730s, but even when the hard times of the 1730s receded, the poor remained. After 1743, poor taxes in Chester County could be increased up to three pence per pound taxed and nine shillings per head for each male; they remained at that level to the end of the century. In addition to providing for the dependent poor, overseers determined eligibility for relief, apprenticed poor children to local farm families, and provided employment for the working poor. The records of townships and overseers, then, are one way of glimpsing a community that was important to those women outside the security of family and cultural communities, particularly outside the Quakers who had the best private relief organization. The women who appear in the records of the overseers of the poor were part of a geographical community not congruent with the Quaker meeting, and therefore the picture that emerges from these records offers another view of women in the Brandywine Valley. Although women were as invisible at the township level as at the county and provincial levels in terms of public officeholding and taxpaying, they were clearly visible as members of a geographic community entitled to public support.[3]

Pennsylvania law provided in meticulous detail for the relief of the poor. In 1750, the law defined who was an inhabitant (anyone who was indentured, employed, paid taxes for one year, or held a lease) and how to provide for those who were not (they must give security that they were self-sufficient or be returned to their own township). Anyone receiving relief had to wear a red or blue badge marked "P" on the right upper sleeve along with the first letter of the county or city. The penalty for not wearing this badge was the loss of relief, imprisonment, whipping, or hard labor. This provision was not repealed until 1771, and Delaware law had similar provisions.[4]

Although no longer forced to wear the badge of poverty after 1771, under the new law rural women could be forced to work in a house of employment. Poor children could be bound out at any age. The law also provided that married women and widows take their place of residence from that of their husband, that parents and grandparents, if able, maintain their children and grandchildren or be fined, and that the property of a husband who deserted be attached until he provided support or was jailed. The new law reflected a greater effort to make the poor and their kin pay their own way, but is also provided greater anonymity for the poor. The only badge of poverty between 1771 and 1798, when the first Chester County poorhouse was established, was the knowledge that a person was receiving support from the community. The story of how townships implemented the law can be traced in a few remaining township books for Chester County.[5]

The case of Heatheren Barkley is a good place to begin that story. On April 22, 1785, the justices of the peace for Chester County requested the overseers of the poor of London Grove to remove her from the township of Kennett. Barkley, the justices wrote, "a Single woman now Bigg with Child and Likely to become Chargable is Come into the Said Township of Kennett." She had been questioned by the justices to determine her legal place of residence. Barkley apparently told them that she had resided in New Garden for at least a year, probably working on one of the farms. The justices therefore determined that New Garden and not Kennett Township was responsible for her should she need financial assistance in birthing and caring for her child.[6]

Unfortunately, the poor books for Kennett and New Garden do not exist, so Heatheren Barkley's experience before and after the decision cannot be determined. The experience of almost a hundred other women in Chester County between the 1770s and 1790s can be traced, however, and a pattern of community for these women can be discerned. One can get some idea of who they were and what their experiences with the geography of poverty was.

The purpose of the residency requirement was, of course, to control relief in a period that assumed everyone had a home township but that people, especially pregnant women or ill persons, might need relief when they were away from it. Older people usually were known in their community and so had no difficulty obtaining relief. Young people moving from job to job, who became injured or ill, had to be questioned to determine where they belonged. Township officials contacted the proper officials in other townships asking them to remove the individual. The person's home township then sent someone to convey the person back. Heatheren Barkley was removed in this way. So too was Mary Furnis and her child, who the overseers of Birmingham Township determined belonged in Christiana Hundred, and Jane Bigart, whom Kennett overseers asked New Garden to remove in 1795. One was a "stranger" if from another township and these "Strang" women were transported out.[7]

Elizabeth Beaty, who was questioned by the Philadelphia overseers of the poor and returned to West Fallowfield in 1773, is a good example of a "Strang woman" removed from one township to another. Elizabeth arrived in New Castle on the *Prince George* from Ireland in 1762. She was indentured to Arthur Berk in West Fallowfield for four years, no doubt to pay for her passage. After one year of work, Berk sold her to his brother-in-law Robert Hoop for whom she served the three remaining years. Then she worked in the same neighborhood for three more years, moved on to Brandywine Township, then to Lancaster, back to Brandywine, and on to Philadelphia. There the overseers judged her to be the responsibility of West Fallowfield, where she had lived for the first year after arriving, and requested the West Fallowfield overseers to convey her and her child to their township.[8]

Once women like Elizabeth Beaty were within their community of legal residence, the overseers had several methods of providing for their care. If the women was pregnant or had just given birth, the overseers tried to find the father to make him assume responsibility. One West Fallowfield entry recorded a charge for bringing Dinah Webb from New Garden and the "persute" of the father of her child, and another charge for marrying them. Or, if the father would not marry, he could be forced to post bond guaranteeing that the mother and child would not become public charges. Jacob Leamy had to post a three-hundred-pound bond for two daughters "begotten on the body" of Ruth Meredith. If the father of the child could not be found, then the father of the pregnant woman might be forced to put up the one-hundred to three-hundred-pound bond. In 1773, Archibald Guy signed a bond for two hundred pounds for his daughter Elizabeth Guy.[9]

If neither father could be found, then the township had to pay for the board and lying-in of the mother. These costs for lying-in, the period in which the mother could not work and had to be cared for during the birth of her child, can be found in most of the Chester County township records that have survived. Costs could run from two to three pounds, a considerable expense. In these cases, however, the expense was short term, for the mother returned to work shortly after the birth of the child. The fees, like the payment for burial of poor women or children, were one-time payments that did not cause a great expense to the community. In the case of one burial, a winding sheet, a coffin, a charge for digging a grave and "hauling of her," plus six quarts of rum for the neighbors was all the West Fallowfield overseers paid out for Sophia Robinson. By selling her goods, mainly clothing, the overseers recovered most of the cost.[10]

When both mother and child lived, the community might have to support them for some time if the mother was too debilitated to work immediately after birth. When the mother but not the child died, the community then had an orphan to care for until she or he was old enough to apprentice out. This might cost twelve pounds a year as it did the West Bradford Township for keeping Katherine Scott's child after her death. In the eighteenth century, daughters of poor women or orphans were usually bound out to neighboring farmers until they were eighteen years old.[11]

Few records of early indentures of rural children remain, for the overseers did not keep systematic accounts. One book for recording assignments of servants does exist from East Caln Township. According to these indentures, most young women were bound out by their parents or by overseers when they reached five or six years of age.[12]

If the new mother could return to work, she could take the child with her and the problem was solved for the community. But the solution was not always so easy. Sometimes the woman—whether for reasons of age or of disability—was carried on the relief rolls for years. Mary Kimber of West Bradford received relief from 1769 to 1776. Catharine Buffington received aid from 1795 until the new county poorhouse opened in 1801 when she was transferred there. Usually, these women boarded with families whom the township reimbursed. Since the amounts varied, some women may have performed services in return for part of their board. Betsy Bishop of East Caln received fourteen pounds a year for keeping Blind Betty in 1795. Others received less for their boarding services.[13]

Because children had a legal obligation to care for their parents, just as parents were legally obligated to care for their children, overseers usually went to children first before assuming the responsibility of relief. Relatives might, however, ask for some assistance in keeping aged kin or aged widowers or widows. After one daughter refused to care for him any longer, William Baily petitioned the court to grant him support so he could live with another daughter. Widow Catherine Woodrow, left aged and helpless after she had expended her estate in her own support, also asked for relief through the executors of her husband's estate. Jane Buller likewise found she could not subsist on an annuity left by her husband, and she had to petition the court to compel his estate to support her.[14]

Compulsion could solve the problem for those with kin who had resources, but a family with poor children or disabled members might need the support of the township. George Martin petitioned for help to support his mentally handicapped daughter Ann because he and his wife were old, had rheumatism, and only seventy acres of land. The township apparently granted him some support but made him put up a bond for half of his seventy acres to ensure that the township would not become entirely responsible for her care. Sometimes the township paid for clothing or helped with temporary relief so the person could be maintained at home. West Fallowfield, for example, bought a cow and feed for Widow Watson in 1793.[15]

By the end of the eighteenth century, the care of the poor was becoming a burden for several townships though not for all. West Whiteland, for example, simply sent Mary Bean "from House to House." Anyone who did not wish to contribute their time to her care could pay thirteen pence per day, but the township as a whole expended very small sums. In 1796, after taking care of Mary Bean and paying expenses of only ten pounds in the 1780s, the township recorded no poor at all. Goshen Township, on the other hand, was paying out 150 pounds by 1797 and Bradford almost ninety pounds by 1799.[16]

The cost to townships in fulfilling their obligations to dependent women had become heavy enough that in 1798 Chester County assumed responsibility by building a county poorhouse in West Bradford. The county was one of the first in Pennsylvania to establish a poorhouse after the state passed enabling legislation in 1798. Later, the movement spread across the state and the nation, establishing the principle of central housing of all dependent poor for whom the community had an obligation to provide support. Cities had already built almshouses earlier in the eighteenth century—Philadelphia's dated from mid-century. New Castle County, too, had built a three-story almshouse on the west side of Wilmington in the late 1790s. The location of the Chester County Poorhouse in a wholly rural area and the elaborate records maintained by it make it an excellent window into the lives of dependent rural women in the early nineteenth century.[17]

Chester County directors of the poor purchased a 150-acre farm with the idea of putting all able poor to work and selling any surplus products for income. The farm included a dairy and a supply of spinning wheels so that some women would be able to pay at least part of their keep. Although there is little remaining literature to indicate the ideology behind the county's action, the primary motive apparently was economic: a central place could provide care much more cheaply than could individuals. The poorhouse was, in fact, a poor farm, where rural people were expected to feel at home and be able to use their traditional skills. There was, too, a feeling that the county in making this collective commitment to its poor was providing them with better care. The records of the directors indicate they were very proud of the farm.[18]

The transfer of poor women and men from Chester County townships to the farm in November 1800 gives us the first systematic, if partial, count of poor women from all the townships. Forty-four women and forty men entered the house. Among the women, only one was Native American, Indian Hannah, and one Afro-American, Black Phillis. The other forty-two were Euro-American. Despite the considerable number of young women with children on community rolls, the women who entered county care were primarily the aged who had no one to care for them and for whom townships had been paying room and board to individual families. Half of the women were between the ages of 65 and 103, another sixteen were between 25 and 64, and only six were under 25.[19]

The new county poorhouse on the West Bradford farm gave material form to the concept of institutionalized community care. The main building was one hundred feet long by thirty-six feet wide and two stories high, with a cellar for storage and kitchen and a third-floor garret lighted by dormer windows. Six-foot passages ran the length of the building with nine-foot, eight-inch ceilings. On the south side of the main floor, an eighteen-foot area was divided into dining room and apartments for the steward and his family. On the north side was a twelve-foot-wide area that functioned as larder and a great hall about twenty-five feet long where inmates—particularly males—could congregate. Women, apparently, were expected to spend most of their time in the kitchen or in their second-floor apartments. The men had a common sleeping room in the garret, seventy feet long and twelve feet wide. Eventually, as we shall see, strict gender segregation was imposed on all but married people, but at first a fair amount of freedom was allowed the inmates, including room visitation. There was, however, little heat, for only five stoves were purchased for the entire building. An outside oven provided space for the baking of bread.[20]

Planning and furnishing the poor farm occupied almost two years from the time the directors purchased the 350-acre farm from widow Deborah Harlan in 1798. Late in 1800, the directors began to purchase farm and household equipment. The day book of the directors of the poor records all this activity in careful detail. They bought and hauled from Johanna Furnace, a nearby ironworks, pots, bake plates, flat and round-bottomed skillets, and iron teakettles. From elsewhere, not specified in the records, they purchased coffeepots and pudding pans, earthen pots and chamber pots, cups, saucers, spoons, and small-tooth combs. The five stoves, each with nine-foot pipes, were installed. There were twenty-seven beds and bolsters, though only sixteen pairs of sheets and twenty-four chairs. The directors must have expected only about twenty-five inmates for the new house judging by these early purchases.[21]

Gradually, the house filled with furniture, food, and clothing. Candlesticks were acquired as well as one street lamp and three patent lamps with fifty-nine gallons of oil to fuel them. It was expected that the farm itself would provide most of the food, so purchases were few—sugar, molasses, pepper, coffee, tea, allspice, cheese, chocolate, salt, and raisins plus a supply of pork, apples, peaches, rye, brandy, wine, snuff, and tobacco. Cloth, men's hose, a shawl, coat and vest buttons, three dozen wool hats and bindings, one hundred needles and thread, and papers of pins were purchased. The women would provide both the skills to finish the clothes and the yarn to be made into cloth. Four spinning wheels and six pairs of cards provided the necessary wool and flax–processing equipment. Later there were entries for large amounts of yarn spun by inmates. The farm also planned to provide additional work for women through its dairy and eight milk cows and acquired a barrel churn and milk pans. The household in time filled with sweeping brushes, rolling pins, lard vats, glasses, and the essential dough trough, and in the wash house there were "putting up pins" and an ash hopper. Caleb Townsend and his wife were hired to run the farm and household for $380 a year, and the poorhouse was complete.[22]

The farm was modeled on those that surrounded it. There were more people and more equipment but essentially little change in the mode of production. Caleb supervised the traditionally male part of the farm household production, his wife the traditionally female. Women inmates contributed as they were able to running the household, providing services, producing textiles, and processing food. It was in effect an extended family, not a family that reproduced itself in terms of farm labor, but one that was managed and operated as a working farm household in which inmates paid for much of their own maintenance. Farm surplus was to be sold and extra labor or professional services (medical and legal) were to be hired, as with any other farm.[23]

The Chester County Poorhouse was the community's collective response to providing for those whom no one else could help. The farm relieved the townships of the growing burden of caring for these people. Here the aged poor, especially, could live in homelike surroundings if not in individual homes. It was not unlike the economic structure of the Shaker families and other communities that were being established with a voluntary religious foundation. This, however, was a secular institution, with an obligation to serve a racially and religiously mixed geographical community.

The county community welfare program in this form lasted less than fifty years. By 1850, concepts of care for the dependent poor had already changed drastically. The idea that multipurpose institutions modeled on the home could adequately serve the community had disappeared. Modest functional buildings run by the inmates and a minimal staff with non-resident professionals being contracted for services were outmoded. Instead public officials envisioned specialized state institutions, designed like large estates with parks, specialized staffs, and a passive, carefully controlled inmate population. The mentally ill were the first to be withdrawn from functional local units. The reforms generated by Dorothea Dix in the mid-nineteenth century led to greater specialization and, although this was not her goal, to greater control and management of poor women.[24]

The growing support for change can be seen in the reports of the Chester County Poorhouse Visitors, those who reported annually on the condition of the house. The buildings were not good for their purposes, they reported in 1845, and there were no shade trees. Nearly 20 percent of the 153 paupers were listed as "insane" or "imbeciles," and these were soon to be removed to the new three-hundred-acre state insane asylum. That move marked the final stage in the evolution of welfare institutions in Chester County during the nineteenth century from township to county to state care. Those who had no family or church care were segregated and removed from the population to a central location, where they were supervised by specialized state-employed caretakers. Orphans, unwed mothers, and the ill were all eventually similarly segregated into specialized institutions.[25]

Such institutions were not necessarily less responsive to the needs of these women, nor was the poorhouse necessarily less responsive than the township system. Individual women caring for indigent women might have done their job well or poorly, with kindness or cruelty, as might individual stewards at the poorhouse or caretakers of the later state institutions. But total institutionalization did isolate the dependent poor from their kin, community, and normal life. In fact, isolation became a principle for those state asylums that housed the insane. It placed them in a highly controlled and regulated environment with little opportunity for physical and mental freedom. Such people became, if not invisible as a group, invisible as individuals. Society could concern itself for their total welfare but forget about their individual existence.[26]

On January 25, 1825, the steward of the Chester County Poorhouse admitted Cassy Eveson, forty-six, colored, and "deranged," and her daughter Mary, four. He noted in his admissions book: "in before." A month later Cassy and Mary went "on leave," but less than four months later they were back. This time Cassy was five months pregnant, he noted, "by unknown person while in State of mental derangement." Three and a half months later, the directors bound out Mary, now aged five, to a local farmer for thirteen years to learn housewifery. A week later, Cassy delivered her second daughter Malinda. From this brief record one can infer several interpretations of her care. The lives of Cassy, Mary, and Malinda, the responsibility for them, the care, the acceptance they received, their ability to live out satisfying lives within the constraints placed upon them is, in some measure, one way in which to judge this rural society.[27]

The background of rural women like Cassy Eveson who asked for and received the assistance of their communities is difficult to determine from the remaining records. There does exist one examination book for the Chester County directors of the poor for the decade of the 1840s that gives some background on the women who sought help. These twenty-four women were mostly from rural areas, although most were not from Chester County. In their lives, one can see the typical lives of the working poor women. Over two-thirds were Euro-American, over one-half, single, and almost 50 percent, illiterate. Their average age was just under twenty-nine, the youngest being sixteen and the oldest sixty-four. Most were working women in their early twenties. About 20 percent were pregnant or had a young child; almost 30 percent had been ill and temporarily unable to work. Two were insane and the husbands of two had abandoned them.[28]

For most of these women work began early in life. Three were bound out very young, one at two years of age. At least four more began working for room and board between seven and twelve years of age. Four of them mentioned that one or both parents had died while they were young, and kin had been too poor to assist in a number of other cases. Employers were the fathers of their children in two cases; in another he was a drover who had moved to Ohio, in another a bond servant, in another a foundryman. These were women who in the main could care for themselves but who had no family or church to care for them during childbirth or share the care of young dependents. They now belonged to a new community where they had to deal with an institution established to give them temporary assistance but not to provide a substitute for the families or the communities they had lost.

There is evidence that these women made the poorhouse their community, however, and that they tried to re-create within it new social relationships. The response of the directors of the poorhouse was to impose more rigorous constraints on the inmates and, eventually, to simply move the institutional responsibility to the state level where greater isolation allowed more complete discipline. It should be reemphasized that the Chester County Poorhouse, like most rural poorhouses before the 1840s, was not a segregated institution, that black and white, old and young, sane and insane, healthy and diseased, cohabited. They were multipurpose institutions for the temporarily or permanently dependent members of a geographical community. It should also be remembered that the directors of the poor considered these to be charitable institutions for caring for the poor on a temporary basis: the old would soon die, the young would move out into productive labor again. Directors saw the task of the institution, then, as one of caretaking, not rehabilitation. As caretakers their duty was to arrange a shelter that met the physical and health needs of the inmates. In 1822, for example, the official Visitors to the poorhouse reported that there "appears to be not quite one sheet for each pauper's bed, and halfe of those much worn," and they recommended that each pauper be given two sheets. What bothered them most, however, was the lack of gender separation. This became the poorhouse scandal of the 1820s.[29]

The Chester County Poorhouse increased its number of inmates dramatically during the first quarter of the nineteenth century. In the first five years after the house opened, about 75 people were resident, usually half of them women. In 1806, there were 35 women and in 1809, 67 women. By 1822 the number had almost doubled to 131. The number of black women had also increased. Only 1 of the female inmates was black in 1800, and 16 were black in 1822 (see Appendix table 12). The ratio of males to females stayed about the same. Although by 1822 a large number of women and men were also being helped outside the house in their own homes by family and friends, the house must have already been quite crowded when the Visitors inspected it in April of 1822. They recommended that the steward separate the inmates during the night unless they were married. Five years later, the report said that the intermixing of the sexes produced "very bad & demoralizing consequences and much increasez, and will much increaz the expense of the establishment."[30]

A look at the steward's monthly reports for the year 1825 not only gives a good picture of the poorhouse female population but also helps explain what concerned the Visitors. Fifty-seven women entered that year and fifty-seven left, either with or without permission. This would seem to indicate no increase, except that, of the women who entered, fully 30 percent were pregnant and many others brought children. In the course of the year seventeen children were born. A number of females died, some of them children. On average, however, the number of inhabitants was growing through the number of births. Women were using the poorhouse as a lying-in hospital and a refuge for periods of unemployment arising because they were pregnant, had very small children, were sick, or simply could not find a job.[31]

Most bothersome to the Visitors was the fact that since their last visitation, of the twenty children born, eight of them had been conceived in the house itself. Thus, concluded the reporting Visitors, "has the establishment intended by the Bounty of the public as an asylum for the aged Sick decreped and unfortunate poor of our County, who might by disease accident or misfortune become unable to Support themselves become a very Monster which altho its members are Daily diminishing in number by Death, possess the power of procreating and reanimating itself."[32]

In this way, the young disinherited women, who had no family, friends, or church, had begun to use the institution to reproduce their community in the same way that wealthier classes used the law to ensure the safe reproduction of their own families. In some societies, children are seen as the children of the community and are welcomed regardless of whether the birth occurs within or outside a permanent nuclear family. In this society, the state of the mother was legally important enough to form the basis of an ideology of sexual restraint. The reality that a group of women were reproducing not only outside legal and social structures but within an institution established only to deal temporarily with slippage in the system outraged the Visitors. Moreover, although the Visitors' report did not mention it, the steward's reports reveal that this racially integrated institution was reproducing a racially mixed population. Of the females in the house at the beginning of 1825, only two were mulatto. Both were young and soon were bound out. But at least four of the women, two white and two black, gave birth to mulatto children. "The Child is Yellow," the steward reported of one white woman's child. Of another woman's child, "it appears to be the Child of a Black man and she admits it to be so." Both children of the black women conceived in the house had Irish fathers.[33]

In the economic hard times in the 1830s, the poorhouse was the natural refuge for the working poor who began to use it as a way to alleviate seasonal unemployment. The Visitors' report for 1845 complained that in autumn and the beginning of winter, the poor "obtrude themselves into the house to lounge away the winter season." The influx in fall was caused by middle-class farmers who hired only summer help during the hard times. Part-time hired labor was profitable; full-time labor was not. It was easier to keep live-in labor in the summer in barns, old houses, or sheds rather than within their homes. Migrants could be taken in and then dismissed before they became a burden to the family. Many of the poorhouse records into the 1840s are concerned with sorting out the residence of the working poor. Many received aid only because they were born or grew up in one area and were not employed a full year in another township. Although the rule was becoming obsolete because of the growing tendency to employ only part-time summer help, the poorhouses clung to the older legal fiction that a person had a right to assistance only if she worked a full year. In turn, the working poor could then use the old geographical system to compensate for an economic order that refused to recognize geography. Some laissez-faire critics of the poorhouse were already demanding its abolition by the 1840s.[34]

The conditions of the working poor steadily deteriorated during the early nineteenth century as Irish and blacks competed for available jobs. The creation of a black underclass, as Carl Oblinger has pointed out, was a major consequence of the growth of both the Irish and the black ethnic communities in southeastern Pennsylvania during the early nineteenth century. However, it is incorrect to label their state "disreputable poverty" and to say that a "mulatto elite" monopolized domestic service. It was rather a situation wherein the older black population had stratified, with some black and mulatto families moving into a middle class of skilled and educated property holders. The population was continually augmented by immigration of unskilled and less literate blacks from the South, and similarly unskilled and even more illiterate Irish people. Many of these new men and women became dependent poor as they aged without family or church care. And the working poor were often forced to become migrant poor, women like men searching for employment, never able to rent or own property. These people, however economically marginal, created a culture that has to be explained in its own terms. Most used the poorhouses as a way of surviving in a hostile environment.[35]

Considering the hardships of the dependent and working poor, amazingly few resorted to activities defined as criminally deviant by their society. Almost a third of the seven women who were publicly executed in Pennsylvania between 1759 and 1809 was sentenced for infanticide. Three of the seven—Jane Ewing, Elizabeth Wilson, and Black Hannah Miller—came from Chester County. By the turn of the century, public sentiment had already turned against executions of women. Twenty minutes after Elizabeth Wilson went to her death in Chester in 1786, maintaining that not she but the father had murdered their twin infants, a reprieve arrived. Public reaction against Wilson's execution helped change the law that same year. The law had made the concealment of the death of a bastard child a capital offense; the new statute required "probable presumptive proof" that a child had been born alive. After 1794 all capital punishment except for premeditated murder was abolished, but women still went to the gallows. Black Hannah Miller's execution in West Chester in August of 1805, watched by two thousand spectators, passed without public outcry. Four years later, however, the execution of Susannah Cox in Reading, Berks County—a "not so bright" servant girl of twenty-four who went to the gallows in a white dress with wide black ribbons—provoked an attack on the hangman and caused the sentencing judge to resign. She was the last woman executed publicly for infanticide in Pennsylvania and the last woman executed in the state for any crime for fifty years.[36]

Remorseful women had been able to escape execution before 1805, and after that time juries often acquitted women charged with concealing the deaths of their illegitimate children. Two dockets remain from Chester County, from 1804 to 1807 and 1812 to 1816. During these years, four of the thirty-eight women arrested were accused of concealing the deaths of their children. The four women spent a total of 391 days in jail before being brought to trial. The only one actually tried, Mary Mercer, was acquitted. Another two were released on bail and a third was sent to Philadelphia; there is no indication that they were ever tried.[37]

None of the other criminal actions of women could be considered gender related. As with men, the most frequent charge was larceny; six women were so charged. Others were arrested for murder, assault, perjury, debt, or intoxication. Several found themselves charged with being disorderly or vagrant but only one each with arson, theft, keeping a tippling house, refusing to be a witness, or, in the case of servants, running away from their masters. Two black women held in bondage were also arrested for running away. Philis Townsend was liberated on a writ of habeas corpus in 1805. Two years later, however, a man claimed a woman, Hitty, as his property. Under Pennsylvania law, some blacks were held in slavery until 1847.[38]

Other information about female offenders in Chester County comes from a very detailed list of prisoners convicted between 1840 and 1846. Of nine women sent to the county prison during these six years, five were black, which is such a disproportionate number that one must conclude that race put a woman in great jeopardy of going to prison. All the black women were convicted of larceny, and all but one were in their early twenties and born outside of Pennsylvania. The one exception was a fifty-one-year-old widow born in New York. These women were listed as first offenders, temperate, able to read, and all but one in good health. Of the four white female offenders, on the other hand, one was convicted for resisting a constable, one for assault and battery, one for fornication, and one for an attempt to pass a bad check. These women were all native-born Pennsylvanians, equally temperate, all but one in good health, and all able to read and write. The main difference between the black and the white women was age. Middle-aged white females but young black females were likely to be convicted of crimes. Young and old, black and white shared gender-defined jobs in prison—they sewed while the men were taught to weave.[39]

The scattered jail records indicate that few women compared to men spent time in jail or prison. Then as now, violence was not acceptable even as deviant behavior for women. Once the laws relating to infanticide and concealing the death of an illegitimate child were relaxed, women seldom transgressed the law. The exceptions were black women accused of theft. White women who went to jail were seldom accused of theft, and they appear to have been much more deviant in their behavior than blacks. A few chose violence, but their infractions of society's rules were so infrequent that factors other than the commission of a specific act must have led to their incarceration and conviction. The public apparently did not see incarceration as a necessity for most deviant women in early nineteenth-century Chester County. While reformers of the early nineteenth century considered isolation and regimentation to be a way to reform social deviants, only a few women, particularly young black women, were subjected to these restraints: they were peculiarly appropriate for the male deviant. Women were far more likely than men to be judged insane rather than criminal, and because they were judged insane, they ended up in the county poorhouse rather than in the county prison.[40]

Noticeably absent from arrest records of women in Chester County was prostitution, another gender-related crime that in later years would place many women behind bars. Prostitution was not then considered a crime, although it might have been considered a cause of crime. Prostitutes plied their trade openly in Philadelphia and Wilmington in the early 1800s, but no mention of such work exists for the rural areas. Nor did women in the poorhouse give any indication of having venereal diseases, which must have been prevalent among prostitutes. Only two women in Chester County were mentioned as being treated for venereal diseases, and one was a resident of Philadelphia at the time of her treatment.[41]

Women usually worked at more typical jobs and suffered more typical ailments. Record books of diseases treated by the physician at the Chester County Poorhouse in 1841 and 1842, for example, indicate that of 107 females treated, the largest number had diarrhea or dysentery. The physician, who seldom made extended notations, added after seventy-year-old Milly Mayarrow's name "dysentery, would not take medicine, died." The water, most likely, was the cause of such upsets. Diseases of the lungs, with coughs and fevers, were second to intestinal disorders, rheumatism third, and diseases of the liver and bladder fourth. While diarrhea affected all ages, children had also to confront the dangers of measles, whooping cough, and worms. Elizabeth Richardson, three, died from convulsions brought on by worms, Ann Jane Campbell from whooping cough. On the whole, the women appeared to be quite healthy considering the difficulty in quarantining those with contagious diseases and the lack of trained nurses. A female who survived infancy could expect relatively good health until the age of fifty in the eighteenth century, even when institutionalized.[42]

The female children of the poor faced the harsh conditions within the poorhouses, and overseers bound them out as soon as possible. In an era that had no institutional provision for orphans, they usually lived in private homes. But the conditions of that life, although in a household rather than an asylum, were far from easy. Chester County trustees of the poor bound out 322 children in the twenty-three years between 1802 and 1825, 40 percent of them female and 17 percent black. The Euro-American daughters appear to have been mostly orphans, indentured at a mean age of five years. Although one can imagine toddlers of two and three being treated like adopted children, there was a clear expectation that they would soon work. Children with physical or mental disabilities were promptly returned by the indentors, and these were the children who remained dependents, growing up in the poorhouse. County officials expected them to labor, as soon as they were able, in the homes of others.[43]

Afro-American daughters of Chester County were indentured at an older age than Euro-American daughters, a condition the documents do not clearly explain. Perhaps black families took in the younger children, so that only the older ones were bound out, or white indentors may have been reluctant to feed and clothe black children who were not yet able to perform much work. At any rate, the trustees of the poor indentured black daughters, on the average, one and one-half years later than they did the white, the mean age being almost seven years.[44] Another possible reason for later indenture of black daughters may be that they were not orphans but children of poor parents.

The New Castle County indenture books do not show an age difference between black and white. Indenture records exist for 382 females for 1818–39. As in Chester County, about 40 percent of the total number indentured were female. In New Castle County, however, a much larger number were black, 32 percent, with both black and white indentured at a mean age of 9.4 years. It is possible that these figures reflect a rising age at which children were indentured, but there seems to be no discernible change in the age of indenture over the period. In fact, children in northern Delaware seem to have been indentured at a later age. The New Castle County indentures are much more complete than those of Chester County in that they give the names of the parents—usually a single parent—of the young girl and are also signed by that parent. Justices of the peace or trustees of the poor signed over a third of these indentures, but fathers and mothers each signed a fourth. Of the signers whose surnames differed from the young women, most were female names, probably indicating that they had remarried. Kin, guardians, or both parents seldom signed the bonds.[45]

One striking characteristic of the mothers and fathers who signed their daughters into bondage is that many were highly illiterate compared to the average in the population at that time. Almost half of all the parents signed with a mark, over half of them women. Illiteracy was a proxy for poverty. Parents bound their daughters over to strangers because they had reached the age where they could work for room and board and the parents could no longer support them at home. Nearly nine years of hard work, to age eighteen, stretched before these youngsters before they could collect their new suits of clothes, sell their labor freely, and receive wages. Rachel Rowlens of Christiana Hundred, for example, indentured her nine-year-old daughter Sarah to Martha F. Willward in 1828 and signed with her mark the bond that would keep Rachel working without wages for eight years, four months, and eight days; she would receive room, board, lessons in reading and writing, and freedom dues of a suit of clothing. Andrew King, a Mill Creek Hundred laborer signed his daughter Eliza Jane's indenture in 1835, binding her for eight years, ten months, and fourteen days. Mary Reading, an orphan, had her bond signed by the justices of the peace on her ninth birthday in 1828, when Joseph and Narsissa Froth agreed to feed and cloth her and give her two months of schooling for the next nine years.[46]

Still these children seemed mature when compared to examples from Chester County, where Martha Hoopes was indentured to Amos Worthington of Goshen five days after her fifth birthday in 1805 for thirteen years, with two years and two months of education, and Phillis Gromley, "of colour," was indentured at six years, nine months, and ten days to Waters Dewes of East Fallowfield, to serve eleven years, two months, and twenty days in 1814.[47]

Among the black population, at least, there is some evidence that New Castle County parents indentured their daughters reluctantly and were sometimes forced into that action by public officials. The parents of fourteen-year-old Juliett Dickerson tried to post bond to avoid her being bound out, but when the mother, Flora Dickerson, could not give sufficient security, Juliett was forcibly bound by the justices of the peace in 1838.

Black daughters were indentured by the trustees of the poor in only a slightly larger percentage than were white daughters, by black fathers only slightly less, and by black mothers slightly more. The main racial difference was that in 6 percent of the Delaware cases masters or mistresses indentured girls, sometimes for longer than eighteen years, using indenture as a means of manumission. In a few cases, black daughters of eleven and twelve were listed as having indentured themselves. The education of black girls in Delaware was also less carefully provided for, although even some illiterate parents arranged for their daughters to be educated. In a number of cases, the indentors offered money to black parents in lieu of schooling. This seldom happened to white parents.[48]

Few records remain to describe what it was like to be a resident of the New Castle County or Chester County poorhouses in the early nineteenth century. There are, however, some newspaper records that document the deteriorating condition of the Chester County Poorhouse by the mid-nineteenth century. As the main structure became more crowded, outbuildings were added and racial segregation instituted. By 1850, a small poorly built structure east of the main building housed the adult black inmates—both male and female—with the second floor and garret reserved for the insane and diseased. In 1853, the official Visitors described conditions in the space reserved for black Chester County inmates as "deplorable," "unfit for habitation," and crowded with twice the number of inhabitants it should hold. Black children had, however, been recently removed to a new well-ventilated building north of the main building where nurses cared for them. Before that time, children apparently were crowded in with adults. White children were housed in the main building with their parents, usually the mothers. After 1853, a small building west of the main building served as a white schoolhouse, staffed by an inmate teacher. There is no mention of a separate school for black children, and nothing is known of their education.[49]

Beyond racial and gender segregation in the main building, some other separation of inmates occurred in Chester County. A small infirmary housed the sick, the most aged, and the "maniac." An article in 1841 described one woman as being imprisoned for nearly a year in a small dimly lighted room with no furniture, and no heat or ventilation; she was strapped into a wooden chair with her feet fastened by iron fetters to the floor. The prisoner, wrote the Visitor, "held no communion but with the demons of delusion" and was kept in this situation "because she was incapable of complaint." When she was released from the chair a few days later—a symbolic act made legendary in similar cases in Europe and America—she maintained the same position, bent and mute from imprisonment.[50]

To obtain better treatment of such inmates, the state, at the urging of reformers, established the Pennsylvania Hospital and Asylum in 1841. Treatment of the insane had always been a problem because physicians believed insanity to be incurable and lengthy institutionalization was expensive. Although the Friends had established an asylum in the early nineteenth century for their own poor, rural Quakers could not afford the expense of commitment even when charged less than the cost of maintenance. After 1825, however, psychiatry developed as a branch of medicine, and by the time Dorothea Dix rode through western Pennsylvania in 1844 visiting poorhouses and urging that new asylums be built, reform was well underway in Pennsylvania. Yet there is some doubt that women like the one described above were ever transferred to the new spacious grounds of the Pennsylvania asylum. The reformers wanted confirmation of their new theory that segregated care for the insane could bring results. Therefore, they saw to it that only patients who were newly arrived at the poorhouse and were good candidates for recovery were transferred to the new parklike asylum. Those cases considered hopeless remained at the county poorhouses year after year with only minimal care and no hope of improvement. Blacks invariably remained in the poorhouse. In 1853, the cellar of the black inmate house still contained a woman labeled "maniac" enclosed in a four-by-six-foot pen with no light, no bedding, and only one coarse garment. Some insane, as the Visitors reported, "were free to cower around the furnace"; others were fastened to pallets on the mortar floor.[51]

By 1853, even the main building was in disrepair, the roof decayed and leaking, the walls unsound, the floors, partitions, and ceilings broken, and, as the Visitors reported, the windows and doors "beyond repair." The original legislation of 1798 had made no provision for funds to build new structures. The reformers of the late eighteenth century had expected the structure to be maintained and the number of occupants to remain constant. But within fifty years, the building was crowded and beyond repair with no way available to replace it without state legislation. Welfare recipients, except for the insane and then only a limited number selected for high probability of cure, were of low priority in the early nineteenth century. The right of the poor to be cared for by their communities had disintegrated under the economic and social pressures of the transition to a modern commercial economy. The 1830s and 1840s were decades in which public architecture flourished in Chester County. Schools, churches, a new courthouse, and a new horticultural hall all testified to the availability of funds for public institutions. But poorhouse conditions continued to deteriorate.[52]

Who was the average adult female inmate in Chester County and New Castle County in the early nineteenth century? She was most likely white, a native-born American, aged forty-two. But the reason for her institutionalization depended on geography. By 1850, half of the Christiana Hundred inmates were listed as Irish and alcoholic. Chester County black inmates were much older, averaging fifty-seven years, and their old age usually accounted for their presence in the poorhouse. Two-thirds of those from Christiana Hundred in New Castle County had a physical disability, one-third a mental disability. Half of those from Brandywine were listed as "simple-minded" or "insane." By 1850, Chester County listed 64 percent of all women inmates as "deranged," "idiot," or "simple."[53]

How can this high and increasing proportion of insane women be accounted for? The reforms of Dorothea Dix had publicized the wretched conditions of inmates with mental problems. She and hundreds of other reformers had successfully convinced the public in most states to establish costly and carefully designed institutions for the treatment, and expected cure, of the mentally disabled. Yet, at mid-century, the county poorhouses remained crowded with women diagnosed as such.

The presence of these women in the county poorhouses could have several explanations. It is possibly an indication of a period of extreme social stress. Women at this time were expected to provide succor and support for males buffeted by the impersonal forces of an industrializing and commercializing society. Women had only the poorhouse to fall back on when they could not meet the increased stress that economic and social change placed on the home. Moreover, insanity was already becoming a culturally acceptable female alternative to violence in American society. The ethnic differences here seem relevant. Irish women were most often defined as alcoholic, a socially acceptable response to pressure for that ethnic group. Black women most often entered the poorhouses because of old age or physical disability. Native-born Euro-American women were the ethnic group with a disproportionate number of insane.[54] This was not a new phenomenon. In seventeenth-century England the pattern was already evident. As the belief in witchcraft declined, women's deviant actions were increasingly diagnosed by doctors as insanity and in American society the tradition continued. The middle class kept their women in the attic. The county poorhouse was the attic for the rural working poor.[55]

Enclosed in a patriarchal household that did not respond to the disintegrating forces of early nineteenth-century America, these women could not express their heresies to society at large. Early nineteenth-century women novelists and poets frequently depicted these insane women. For some, their treatment became a metaphor for the control society exercised over their own lives. Emily Dickinson expressed the feeling most clearly: "Assent—and you are sane— / Demur—you're straightway dangerous—And handled With a Chain—." The evidence from the Mid-Atlantic poorhouses indicates that this assertion was based on a reality taking form during the early nineteenth century—the predominance of women classified as insane and institutionalized.[56]

Insanity was the most frequent path of young native-born white women to institutionalization in rural society. Insanity then, as now, masked the inability of women to cope with a hostile, unsupportive environment, and the willingness of a society to label their failure as a socially acceptable deviance.

PART II

THE MARKETPLACE

CHAPTER 5

The Economics of the Butter Trade

For the middling families of the Philadelphia hinterland strategies for prosperity were more important than survival strategies. Within this culture, women were active shapers of economic development. Seeking ways to make their farms more profitable and increase their cash income to purchase commodities from the commercializing cities and industrializing river valleys, women shifted their work from textile production to dairying. In the late eighteenth century, farm families diversified their farms; in the early nineteenth century, they specialized. In both periods, women's involvement in the butter trade provided an underpinning for the economic development of the Mid-Atlantic region.

Almost every historian of this area's economy has noticed the economic growth of the Philadelphia hinterland. Alan Tully cites the West Indian and coastal export trade as the most crucial aspect of the growing rural economy from 1726 to 1755. Immigrants flooding into Philadelphia also increased the market for local farm products. Moreover, all economic groups shared to some extent in the accumulation of material possessions that marked this time of growing affluence. While the wealthiest benefited most, the growing agricultural market enabled middling farm families to diversify and to keep pace with the accumulation of wealth.

After 1750, the strategies already adopted were expanded so that farm families could take advantage of a strong demand and increase in prices for agricultural products. Temporary dislocations struck in the late 1770s and 1780s, thus increasing interest in diversification, but the prolonged rural depression did not come until after 1817 when peak prices for farm products collapsed because of the decline in foreign trade. During this period, however, the development of specialized dairy production cushioned the decline in exports and laid the foundation for a flourishing hinterland economy in the 1850s.[1]

Export butter is easiest to trace. No one has yet studied this foreign trade in detail, but the outlines are clearly visible from the statistics available after 1769. By that time colonial Philadelphia had developed a lucrative West Indies butter trade. The city shipped 47,860 pounds of butter to the West Indies between 1770 and 1771. This export trade would continue with minor interruptions through the next fifty years. From 1770 to 1772, Philadelphia also shipped 146,265 pounds of butter through the coastal trade, an indication that butter making in the Philadelphia hinterland had developed sufficiently to send a surplus to other mainland colonies. Such extensive foreign and coastal trade indicates that the butter business developed extremely quickly after 1750, for within twenty years Philadelphians were exporting a surplus of over 200,000 pounds. Demand was so brisk that it decreased the local supply, causing urban citizens to protest. A 1771 petition from Germantown asked that the export trade in butter be controlled because butter was becoming so scarce.[2]

While disrupted somewhat by the Revolution and postwar depression, foreign butter exports from the new nation continued to increase. They rose to 2.5 million pounds in 1796 alone, with three-fourths of it going to the West Indies. By that year, American butter was showing up in China and the East Indies as well. Disruptions in the early nineteenth century caused by foreign wars and restrictive trade policies reduced butter exports after 1812. Still, export statistics showed annual exports of 1,785,936 pounds of butter between 1802 and 1812.[3]

By the time the export trade declined, the domestic trade in butter had a good start. The growing domestic demand for butter was certainly an important source of income for farm families and was to prove more permanent than export demand. The urban population of Philadelphia was the most important consumer group for the farmers of the hinterland, and it is possible to use it as a crude index for one part of the domestic butter trade. A rough way of estimating Philadelphia demand for butter is to estimate the amount of butter consumed per person and the percentage of the urban population that had cows and so could produce its own butter, and then figure the total amount of butter purchased.

The first question to answer is how many urban families produced their own butter. Although there is no detailed study on the decline of urban dairy production, Billy G. Smith has estimated that as early as 1771 only 3 percent of the households of the free male workers he studied in Philadelphia had cows. Poor people seldom could afford cows and the wealthy minority usually kept them only for their own use. During the war when urban food supplies ran low, many families slaughtered cows for meat. After the war, few urban families replaced their cows, as country butter became readily available and health ordinances increasingly prohibited the keeping of animals in urban areas. Thus, almost the entire population of Philadelphia probably purchased its butter between 1770 and 1850.[4]

The population of Philadelphia grew prodigiously during these years. According to recent estimates, it had passed 30,000 by 1772. By 1790, it had topped 42,000. Thereafter, the percentage of increase dropped somewhat, but by 1850, over 121,000 people lived in Philadelphia. How much butter did the average person consume by 1850? Smith estimated 13 pounds a year for his 1771 male workers. Twentieth-century experts have used the figure of 15 pounds per person per year as their average. Earlier impressionistic estimates by mid-nineteenth-century experts, however, gave 25 pounds per person as the standard for 1850. Perhaps, then, one might arrive at 13 pounds as the minimum. Using this figure, one arrives at 350,000 pounds of butter for Philadelphia in 1771, 546,000 in 1790, and over 1.5 million pounds by 1850.[5]

Although only estimates, these figures do indicate that an extensive butter trade developed quickly after 1750. Farm families produced increasing amounts of butter for the market from that time until 1840 when the national government first attempted to collect and record the total pounds of butter produced on American farms. One wishes this early trade in butter were better documented. The best evidence for budding commercial interest is in the experiments of Charles Read, the eighteenth-century New Jersey butter enthusiast who in the 1750s reported neighbors producing 200 to 500 pounds of butter a season along with great quantities of cheese. By the 1760s and 1770s, there are records of small quantities (up to 22 pounds) of butter being sold and traded by women in western Chester County. Quaker teacher and farmer Benjamin Hawley recorded large quantities of butter purchased for cash from neighboring women. In six months of 1763, he purchased 170 pounds from neighbors Ann Dunkin, Catharine Dilworth, Mary Darlington, Susanna Davis, Lydia Woodrow, Ester Martin, Deborah Taylor, and Hannah Woodward.[6]

Such accounts are tantalizing, but there are few of them. Apparently the women seldom kept accounts of their butter sales because they had not received training in bookkeeping, and men tended to record only their own transactions. A young Chester County boy painfully copied down in his 1770 arithmetic book some rules that very well might explain the process at that time. "Barter or truck," he wrote, "is no more than Exchange.... Suppose A has 144 Ells of Linen Cloth... Which he would truck With B for Butter 100 lb. how many pounds of Butter must [unreadable] for the Linen." Although probably few women were trading butter in so large an amount, the accounts of men show that they were already selling large quantities of butter for cash as well as trade well before the Revolution.[7]

The Revolution disrupted a brisk butter trade in colonial Philadelphia as well as its export trade. Butter and beef had been two of the commodities colonials were accustomed to purchasing in specie, and the war caused inflation and disappearance of specie. Rural areas maintained their old commodity prices, however, continuing to exchange butter at the old rate. Thus farmers maintained trade in butter during the war by permitting consumers to continue their normal demand. According to Anne Bezanson, farmers then profited from an active market in butter while foregoing the use of imported commodities.[8]

The main disruption of the butter market seems to have occurred when Howe and Cornwallis swept through Chester County in the fall of 1777, leaving several hundred angry farm families in their wake, many of whom later filed claims for destruction by the British. According to these claims, soldiers confiscated over one hundred cows, including herds of sixteen, eleven, and ten cows, as well as churns, milk pans, pails, skimming dish ladles, tubs of butter, and thousands of pounds of cheese. Some farmers were very precise in their claims. Thomas Levis, who lost only one cow, wrote: "30 lbs. of buter and as mutch Cream as would have made 10 lb of buter & broke the Pots." Patrick Anderson who lost eleven cows wrote: "One Barrel Churn Stove to pieces." Quakers submitted accounts of lost cows to their meetings as well as claims for numerous cows and heifers along with other farm products surrendered to the American officials in fines for refusing to pay war taxes and not attending muster. It was this 1777 disruption that led urban diarist Elizabeth Drinker to record a butter famine in Philadelphia and Sarah Fisher to report country friends outside the lines sometimes smuggling in a few pounds of butter for their use. Even after country people could again hold markets in Philadelphia, Elizabeth Drinker worried that they would not come unless they could get goods for their produce. Merchants in 1778 had little to trade.[9]

With the war over, however, road building commenced, trade revived, and the completion of the Philadelphia-Lancaster turnpike in 1794 brought country folk into Philadelphia with their butter once more. By the 1790s, some farm women in western Chester County were preparing large quantities of butter for the Philadelphia market. Samuel Taylor recorded in his 1798 journal: "My mother took 40 lbs. of Butter to Gedion Williams. Sold it for 1/6 per lb." In June 1799 he recorded that he helped start his sister Deborah on a two-day trip to the Philadelphia market with 30 pounds of butter. In August, Deborah and Samuel both took butter to the Philadelphia market, stayed overnight, and returned the next day after selling their butter for one shilling, ten and a half pence a pound.

Later that same month, Samuel recorded sales of 32 pounds of butter in Philadelphia, and on another occasion that year they sold 35 pounds. Unfortunately, Taylor's diary record covers only a little over a year; yet it appears that the Taylor women produced about 170 pounds of butter for sale during that time.[10]

The early nineteenth century is strangely devoid of butter records. The records no doubt exist, but they have not yet surfaced in historical collections. The accounts that are available show that the quantities sold by individual farms increased. When the du Ponts began to buy butter in 1815, probably for workmen boarded at the new gunpowder mills, the Gregg farm supplied 235 pounds of butter during a four-month period from May to August. The following year the farm supplied 188 pounds in a period of less than four months. The prospects for farms selling produce to the new factories being built along the Brandywine seemed so good that farmers in the area petitioned Congress in 1816 to raise protective barriers to "enable the manufactories to continue their works & thus give employment to the helpless & indigent & furnish us with a home market for our products."[11]

While larger farms looked for sales of hundreds of pounds, smaller farms and tenant farmers were content to sell a small surplus. Patrick Kenny, a Catholic priest who kept a farm not far from Wilmington in the early 1820s, recorded almost weekly trips to market with 6 to 20 pounds of butter produced by his tenants. On February 6, 1822, he recorded Eliza's first marketing trip, taking 4 pounds of butter to market and returning the following day with eighty cents.[12]

Once the butter trade had been established, farm families nurtured it because it was such a secure source of income. The high and relatively stable price for butter between 1785 and 1821 provided a steady incentive for farm women to devote an increasing amount of their time to processing and marketing the product. Wholesale prices for firkins of butter in Philadelphia remained over ten cents a pound except for 1788–89, 1792, and 1802–03, never dropped below seven cents even during the price volatility of the 1820s, and reached a low of five cents only in the deflationary period of 1843. Prices were over twenty cents during some months in 1805 and 1817 (see Appendix table 13).[13]

Butter sold locally in southeast Pennsylvania for six to ten pence per pound in the 1760s and 1770s and between one shilling, eight pence, and one shilling, ten pence, in the 1790s. After the new monetary currency was established in the first decade of the nineteenth century, it sold for twenty to twenty-five cents a pound in 1815 to 1821 and seventeen to twenty-two cents in 1845 to 1850. The sale of a surplus of several hundred pounds a year was not a small item in a farm budget. It was often enough to buy most of the commodities the family needed for the household.[14]

Two detailed accounts kept by farm women exist for the 1820s and early 1830s, those of Martha Ogle Forman in northern Maryland and Esther Lewis in Chester County. The contrast between the two women is important because one woman was born in northern Delaware, married a man from Maryland, and produced butter on a slave plantation; the other was a Quaker widow who inherited her husband's farm on his death. Martha Forman's husband owned about 800 acres of land among all his properties of which the Rose Hills farm for which Martha left a record was only one. At his death, she was left with an ample stipend but none of the farmlands. Esther Lewis became a widow in 1824, fought to retain control of the 114-acre family farm when relatives objected to her controlling the property in her own right, discovered and mined iron ore on the property, managed the farm, cared for an aging mother and dependent kin, and eventually divided the farm equally among her four daughters. Despite the great differences between these two women, each with the aid of servants processed thousands of pounds of butter during the 1820s and 1830s.[15]

Martha Ogle Forman began her entries in 1816 when she sent butter to Baltimore for twenty-five cents a pound. She was also selling 35-pound crocks to neighbors and trading large quantities for food from them. On June 1824, she recorded: "We now make more butter than we know what to do with." Three years later, when she was gone from July through September, her black dairymaid sold 215 pounds of butter. Soon after, she recorded making 25 pounds a week and in July 1830 wrote: "I have churned 52 lb of butter this week."[16]

Esther Lewis, who began her diary in 1830, churned at least 288 pounds of butter that year and sold about a third of it in Philadelphia and to local people. The following year, after adding another cow to her herd of three, she churned 338 pounds in five months with the help of a servant. Of this amount, she recorded selling a little over a third. In addition, she recorded large quantities of butter made and sold by the tenant woman living on the farm with her and by one other woman living nearby. By 1835 and 1836, Lewis was recording monthly trips to Philadelphia, carrying the joint products of three households, which ranged from 75 to 100 pounds each month.[17]

During the 1840s, production increased, although few accounts precisely document the increase at the level of the individual farm. The Strawn butter and egg book, kept by a farm family in Bucks County from 1845 to 1850, gives some idea of the way in which butter sales mounted during the 1840s. According to the census of 1850, Margaret and Eli Strawn were both twenty-eight and had two small children, Henry, four, and Mary, two. Two young people, apparently hired hands, Samuel Teornd, fifteen, and Caroline Shelly, seventeen, lived with them. The Strawns owned sixty-five acres of land, seven cows, and a diversified farm that produced wheat, rye, corn, oats, potatoes, and hay. Their butter book shows the following amounts sold during the 1840s:

1845	347.5 pounds
1846	320.5 pounds
1847	449.5 pounds
1848	461.5 pounds
1849	686 pounds
1850	864 pounds

Numerous farms in Pennsylvania and Delaware must have increased their output in similarly impressive ways, as they switched to cows that produced more milk for longer periods, purchased more efficient butter churns and processing equipment, and, no doubt, worked harder.[18]

Statistics issued by the United States government from 1840 to 1850 reflect the commercial transformation of butter making. According to the figures for 1840, Americans had already developed an extensive trade in butter. Dairying had moved from a peripheral part of farm work to become one of the leading agricultural pursuits in the country. By the late 1840s, canals and waterways were full of ships carrying butter in kegs and tubs. New York farm families were selling millions of pounds of butter yearly, to places as far away as New Orleans. The lake port of Buffalo imported over 100,000 kegs of butter for transshipment in 1849. Cleveland alone imported over 2 million pounds of butter that year.[19]

The Mid-Atlantic states quickly took the lead in butter production. By 1840, dairy products ranked fifth in value among agricultural products of the four Mid-Atlantic states and most valuable among animal products. Mid-Atlantic farms were processing over $15 million worth of dairy products a year. A butter belt surrounding Philadelphia was already visible. Eleven counties in three states were part of that belt: Delaware, Montgomery, Bucks, and Chester counties in Pennsylvania; New Castle County in Delaware; and Sussex, Burlington, Gloucester, Salem, Cape May, and Monmouth in New Jersey. Similar butter belts surrounded other major cities.[20]

The segment of the Philadelphia butter belt that linked Delaware, Chester, and New Castle counties can be examined in detail through the 1850 agricultural census. In that year, 4,760 farms in these three counties produced over 4 million pounds of butter. From the census, one can also estimate how much surplus butter was available for the market (see Appendix table 14).

How much surplus butter women had to sell by 1850 depended upon how much the family could be expected to consume. Economist Fred Bateman has argued that the traditional amount used by twentieth-century economists, 15 pounds, should be increased by 70 percent for farm families, which would bring the figure to about 25 pounds per person.

Since nineteenth-century experts used this higher figure even for urban areas, it seems safe to use Bateman's estimate as a minimum. Two hundred pounds of butter could then be expected to provide for a household of up to eight people. By dividing farms for the three counties of Chester, Delaware, and New Castle into five categories, we can arrive at an estimate of how much surplus butter was available. For purposes of estimating location and amount of available butter for sale, I have divided the categories into (1) self-use, 0 to 199 pounds, (2) surplus, 200–599 pounds, (3) middling dairy, 600–1,999 pounds, (4) large dairy, 2000 pounds and over, and (5) commercial dairy, 8,000 pounds and over (see Appendix map 1). Using this division, about 25 percent of the farms produced for use, 50 percent had a surplus, and another 25 percent produced amounts ranging from 600 to 1,999 pounds. Only 8 percent could be categorized as commercial dairies. These largest dairies produced about 2 percent of all the butter, the middling dairies produced 27 percent, the surplus farms produced 44 percent, and the self-use farms produced 25 percent. Thus, the farms that produced a surplus accounted for half the total amount of butter produced.[21]

In 1850, a Delaware farm reporter boasted that dairies of fifteen to one hundred cows were common in New Castle County. He exaggerated. There were no dairies of one hundred cows in the 1850 census. Large commercial dairies were common in late eighteenth-century England where they provided the large urban population with butter, and similar dairies had long been a dream of agricultural visionaries in Delaware. John Spurrier, who moved to Brandywine Hundred in New Castle County in the early 1790s, advised his new countrymen to emulate the British. He thought that dairies of one hundred cows each could be maintained on plantations of thirty acres of improved grass with two men to clean the barns and feed the cows, four dairymaids to milk and make butter and cheese, and two dogs to help power the churns. A half century later, Philip Reybold attempted to implement Spurrier's plan in Delaware. Reybold, a relatively poor man when the Chesapeake and Delaware Canal opened across the isthmus of Delaware in 1829, built up a herd of almost one hundred cows within the next six years. Edward Canby, the Brandywine miller and gentleman farmer who visited him in 1835, recorded in his diary: "He keeps about 100 cows, sends his butter to Baltimore where it is eagerly sought for at good prices. English family fully employed." By 1850, however, Reybold had divided his farm among four sons and no one had more than sixty cows. They had diversified by growing peaches and selling peach trees.[22]

There were, however, 6 farms in the three counties that produced a total of almost 75,000 pounds of butter in 1850. Another 409 farms produced 2,000 pounds or more each, and 1,613 farms produced between 650 and 2,000 pounds. These 2,028 farms, 25 percent of all the farms in the three counties, produced large surpluses. Another 50 percent probably produced surpluses ranging up to 500 pounds a year. These farms were arranged in a butter belt surrounding Philadelphia, with the highest per capita farm production in the eastern part nearest the city and lower per capita farm production in the western part (see Appendix table 15).

The agricultural census of 1850 allows us to see these farms in microcosm. Brandywine Hundred farms, where this study began, is a good place to view an average farm in the final stage of the butter trade in the century from 1750 to 1850. The average Brandywine farm was worth $63 an acre. At eighty-seven acres, this amounted to over $5,000 a farm. Almost 80 percent of the land was improved, and the farm had $234 in machinery and $386 in stock. It produced over $100 in slaughtered goods and market and orchard produce. A farm this size could produce 170 bushels of corn, 100 bushels of wheat, 80 bushels of oats, over 60 bushels of Irish potatoes, and almost 20 tons of hay. None of the farms produced wool, flax, or rye—all products that normally required female labor in earlier periods. More important, for our purposes, 75 percent of the farms owned milk cows, averaging over six per farm, and produced 630 pounds of butter per farm. At a retail price of twenty cents a pound, this amount could yield $125 a year, an important cash income for a farm family when the average farm had less than twice that amount invested in machinery (see Appendix table 16).[23]

Women could produce such huge quantities of butter primarily because of the increase in their skill. They now had to prepare a product to meet competition in the marketplace. Bad butter would not sell or had to go for lower prices. A skilled butter maker in the family meant not only an important cash income but also a steady one, and one that families felt they could predict with relative certainty. There seemed always to be a market for quality butter. Women in the Philadelphia butter belt had developed a cottage industry that was finely tuned to the market and in which they often participated at every stage—milking cows, making the butter, riding the wagons to market, and selling their product from the curb of Philadelphia or Wilmington market streets.

All this work, of course, took increased time from their more traditional duties, especially the preparation of cloth. By 1810, Chester County farm families had specialized in textiles and were producing large quantities of yardage in small factories and shops. In a federal census that year, the county ranked third and seventh among counties producing linen and wool in Pennsylvania, accounting for over 170,000 yards of linen and almost 75,000 yards of wool. By this time, however, Philadelphia and New Castle County had already started to manufacture small quantities of woolen and cotton cloth in larger factories, and the yardage would soon be available in quantities and at a price that made it more profitable for women to devote their time to butter making. The presence of an alternative market product probably hastened the transition to store-bought cloth during the 1830s and 1840s. Butter making was also less tedious than textile processing and offered more variety than either spinning or weaving. It seems likely that the production of butter also speeded the transition to other store-bought commodities. The equipment needed for butter making required low capital investment in tools, which, when combined with the skills women could develop, brought an income that financed the purchase of other commodities previously produced at home. Farm women could afford to use their time in more productive ways now.[24]

The women also turned increasingly to new ways of obtaining assistance in their work. It seems probable that young children were not able to help much with butter processing. Although children might learn to milk efficiently at a young age, perhaps at eight or ten, and might help with the scalding and cleaning of equipment, they could not handle the large milk pans, nor were they necessarily of help in churning. The plunge churn required an up-and-down motion that was very tiring, and it was difficult for a young child to provide the continuous "short, sharp strokes" that advisers recommended. Churning entailed three stages: in the first, the cream was thin and the churner had to churn quickly; in the second, a steady slower rhythm was necessary; and in the third, a slow churning was required. Even if a child might perform the first stage, she would be unlikely to be able to complete the job. A barrel churn, with its rotary crank, would have been easier for a young child to manage, particularly when the churning began, but like turning the crank on an ice-cream churn, it became more difficult when the cream solidified. If a large churn holding twenty pounds of butter was used, a dog or a sheep on a treadmill would be more helpful.[25]

The heaviest period of churning also occurred in spring and summer, times when chores on the farm were heaviest and there were plenty of outdoor jobs for children. In 1830, Esther Lewis noted the tasks her children were performing, as well as those of the women and men on the farm. Her daughters were whitewashing the house, picking cherries and beans, and weeding the garden. They did not assist her with the churning.[26]

The switch from spinning to churning may, in fact, have liberated more young girls to go to school in the winter. Earlier, girls had attended school primarily in summer when the boys were working in the fields. Processing of flax took place mainly in the winter after the harvest, and girls usually were busy at home at the wheel by the time they were seven years old. Since little butter processing was done in the winter, girls were now free to attend school then, and particularly during the period from January through April.

If churning liberated young girls in the winter, it ensured that they would have more to learn from their mothers regarding the processing of butter as they became older. Again Esther Lewis left one of the few accounts to discuss the initiation of daughters into the mysteries of butter making. Lewis had been sending butter from her West Vincent farm to Philadelphia for nearly ten years when she wrote in 1837 that sixteen-year-old Graceanna was now "principal Dairy maid." Five years later, when Graceanna was teaching in York, Pennsylvania, Lewis wrote that she had taught another daughter, Elizabeth, then eighteen, dairy skills: "I had no idea," she wrote proudly, "when I taught her to skim &c that she would have made a dairy maid so soon. She has churned & printed twice, handsomely."[27]

Mothers expected daughters to be able to churn, but they also might hire servants to allow daughters to attend to tasks other than churning. Lewis hired a servant in 1839 because, she said, "I wanted the girls to have more time to get their sewing done." Again in 1841, she noted she had hired an extra girl because "the girls have a great deal of quilting and sewing to do this winter." Thus the woman of the house first trained her daughters and then managed the allocation of their time, for these daughters remained at home after fifteen years of age far more often than did sons. It is difficult to tell how many farm women had servants to help them with their butter-making tasks, but by the late eighteenth century the dairy accounts of middling farmers indicate they expected to have at least one hired girl for the summer and often one year-round with extra help in the summer. Poorer farm families hired out their younger women. By 1850, at least a third of the Kennett farm families had live-in female help, another third used family labor, and the rest furnished labor for other families. Similarly, a third of the farm families in Brandywine and Christiana hundreds in New Castle County had live-in female help.[28]

The relation of employer-servant was always a difficult one, particularly on the farm where work was hard and most women could look forward to managing their own farm households, if only as a tenant farmer. Servants were very scarce in early Pennsylvania, as Deborah Morris recalled in the late eighteenth century: "Few of the first settlers were of the laboring class," she wrote, "and help of that sort was scarcely to be had at any price, so that many of the women set to work they had never known before." She recalled her great-aunt had helped with building, handling one end of the crosscut saw with her husband. Even after the pioneer stage had ended, rural indentured servants were difficult to find, as Sharon Salinger has shown.[29]

The advertisements in the *Philadelphia Gazette* as early as the 1770s indicate that specialized dairymaids were in demand. Some people advertised for women capable of managing dairies or for couples in which the wife had been "used to a dairy." Others advertised the sale of indentured servants with dairying skills. But even this early, there was some evidence of urban women not wishing to go into the country to work. One advertisement for a black bondswoman said: "She will not be sold into the country." By the end of the century, certainly, farm women in the rural areas depended on one another for work. The daughters in laboring families looked to middling farm women to hire them and to provide them with wages and room and board. The middling farm women, in turn, expected the younger daughters in laboring families to hire out to them when they reached twelve to fourteen years of age.[30]

Farm families, rather than indenturing young girls, probably found it easier to offer room and board and a small wage or to hire only seasonal help, and to dismiss the women if they did not meet the local standard of housewifely skill. Because females were indentured only to eighteen while males could be indentured to twenty-one, it was not economical for the indentor. Like male labor, however, female labor was often in short supply. Hired girls may have had little incentive to seek jobs as operatives in the new factories springing up along the Brandywine after 1810 because they could obtain work closer to home at comparable wages. In fact, the demand for women in agricultural labor may have contributed to the steep increase in women's wages in the 1830s.[31]

Although farm accounts clearly indicate a large number of young rural women working for wages on farms, they do not usually describe the type of work they did in detail. Apparently the level of skill expected of local Pennsylvania girls by the 1830s was high, although the farm women themselves may not have realized this unless they had an opportunity to compare local skills with those of women who came from an outside area. Irish immigrants of the 1830s who sought jobs in the country provided one such comparison. Lewis, who normally employed local women to help her with the churning, employed an Irish woman "fresh from the Sod" for a few weeks in the spring of 1839. When she wrote later of the "Irish awkwardness," it was clear that she was describing a young woman from an area where rural households had few material items. "One day," Lewis related, "after assistance and showing till she got the cream in the churn, I told her to put the lid on, she endeavoured to comply and worked it every way she could, as she thought, and could not make it fit. I looked at her a while and then showed how nicely I could make it fit, by turning it upside down." Lewis went on to say that she now had two women who, with her daughter Mariann, were able to provide all the assistance she needed in maintaining the household. Irish women, of course, also soon mastered the skills of butter making, and some eventually were at work on their own farms processing butter.[32]

Black countrywomen participated in the growing butter business as well. Martha Ogle Forman's experience with black dairymaids may be some indication of the level of skill that black women developed in dairying. Forman had two black dairymaids, both bondswomen, one of whom died in 1820 and the second in 1841 after handling all the dairy work for almost twenty years. Forman described the process of learning for Rachel Teger who attained a high skill in butter working in this epitaph: "She was a field hand when I came here but I soon discovered she had a great deal of intelligence and industry and she made all my pasteries, Cake and biscuits, best Candlemaker and dairy maid, she has made me many thousands of butter, she had been dairy maid for upwards of 20 years."[33]

To say that women were busy at the churn is not to say that they were automatically being loosened from the older bonds of the patriarchal family and tied to the marketplace in new liberating ways. Butter making could exist within the older labor structures. Black women in Maryland who produced for the Baltimore butter market remained in slavery. Black women in northern New Castle County, although almost all were free of bondage by the 1850s, also labored in a slave state where few could own farms or produce their own butter for the market. Similarly, many young Euro-American women could not afford farms when they married and were not able as tenants to produce butter to sell. The growing market for butter merely prepared these women for wage labor on farms or in the cities.

For those who lived on farms of about a hundred acres, however, butter making along with other farm enterprises could provide an important way to keep the family together on the farm in a time of increasing land costs and decreasing land ownership. Although the purpose may have been preindustrial, that of providing for the family, the means were commercial. And the work was of major consequence in providing an economic infrastructure for the expansion of industrial capitalism. Farm families, like urban families, could join in the new era of consumption. The difference was that rural women remained producers as well as becoming consumers of the new industrial age. And that was an important distinction.

CHAPTER 6

Churns and Butter-Making Technology

During the century in which butter became a central commodity on Mid-Atlantic farms, the technology of butter making grew elaborate. It left a rich material culture, principally in churns. The study of churns and other butter-related tools thus provides one more way of analyzing the work of rural women. Churns, a part of the material life of our foremothers, are often seen in museums. Whether lovingly jumbled together with other farm implements from the age of wood, decorously displayed as a part of domestic life exhibits, or still in use at living farm museums, the churn is a constant reminder of the important work farm women did in the past. Every museum has at least several churns, but almost every curator has difficulty re-creating the historical context of these rural artifacts. There has been little historical research to explain this rich material culture.

Churns and other butter-making tools are an important material link to butter making, a process by which women extended the old concept of the productive farm household into the new industrial era. Thus the culture of dairying can help provide a context for larger economic models. In 1980, Edward Pessen asked in the pages of *American Historical Review*: "How Different from Each Other Were the Antebellum North and South?" His conclusion was, not very much. For all their distinctiveness, he wrote, "the Old South and North were complementary elements in an American society that was everywhere primarily rural, capitalistic, materialistic, and socially stratified, racially, ethnically, and religiously heterogeneous, and stridently chauvinistic and expansionist." In an *AHR Forum* following Pessen's article, several historians engaged in lively dialogue over his conclusions. None, however, pointed out that Pessen made his conclusions without referring to gender or explicitly to the work of women during the antebellum period. Like Fernand Braudel in his even broader economic analysis of Western capitalism, Pessen seemed to conclude that women's work did not count for much.[1]

Despite the absence of dairying and women's work from historical studies, the history of butter making for the market, in which women played such a crucial part, marked an important stage in capital development. Women's provision of butter both for use at home and for the market economy helped American capitalism develop in a particular way. Butter making enabled large numbers of rural people to participate in the early nineteenth-century American capitalist economy through the purchase of manufactured goods, for as the farm population became more productive for a greater part of the year, their increased income allowed them to buy more consumer goods. When women's work is viewed in this way, it is difficult to omit it from consideration of the antebellum economy.

The butter so actively traded by farm families in the Philadelphia hinterland was not, of course, the product of women's work alone. Women shared dairy work with men, probably along a fairly sharp division of labor by gender. Providing food and shelter for the milk cows was usually the male's responsibility. He built the barns, produced the hay (although women frequently helped him harvest it even after the introduction of scythes with cradles), fed the cows, and cleaned the stables. Before 1800, family, slave, and indentured or kin labor was available. After 1800, indentured labor was joined by hired labor, white or black (see fig. 1).[2]

The job of most farm men stopped with the care and management of cows. At this point in most dairies, farm women, or dairymaids who specialized in butter making, took over the dairy tasks. Women handled most of the milking; there were some male milkers in parts of the United States and England before the 1850s, but it does not seem to have been a common practice in the Mid-Atlantic region. An 1856 dairy essay reported that men in some parts of the United States milked cows, but the author protested that because women were more gentle and clean they should do the milking. Until 1850, however, women certainly performed at least most of the milking chores.[3]

They also usually separated the cream, churned it, worked the butter into solid blocks, added special ingredients to preserve it for long-distance trade, packed it for market, and often took it themselves to trade in Philadelphia or the burgeoning seaport town of Wilmington. Women had the help of younger children, sometimes boys as well as girls, female relatives, or a hired servant, white or black. This division of labor proved to be efficient, for adults could combine their dairy work with other farm chores, arranging their own schedules to fit the daily morning and evening

FIGURE 1. **Farmyard.** From *The Progress of the Dairy: Descriptive of the Making of Butter and Cheese for the Information of Youth* (New York: Samuel Wood, 1819). The Sinclair Hamilton Collection of American Illustrated Books, Princeton University Library.

feeding and milking as well as the heavy seasonal demands of growing feed and processing milk (see fig. 2). After the mid-nineteenth century, when the commercial importance of butter making was more evident to the male farmers in the area because of the competition from western grain and cattle–growing regions, men probably gave greater attention to the processing of butter. As one visitor to a West Chester farm reported in 1867, the establishment was "rather a butter factory than a farm." The visitor went on: "The farmer himself is his own dairymaid, and attends in person to every detail of his dairy." That male "dairymaid" in his farm "factory" perhaps symbolized the beginning of the transition to a new stage of industrial development in the production of butter in the Philadelphia hinterland.

Dairy changes initiated by men seem to have been primarily concerned with the care of animals. Among the early changes that the expanded

FIGURE 2. **Milking.** From *The Progress of the Dairy: Descriptive of the Making of Butter and Cheese for the Information of Youth* (New York: Samuel Wood, 1819). The Sinclair Hamilton Collection of American Illustrated Books, Princeton University Library.

trade in butter brought were improved care and feeding of cows by the men. Farmers began experimenting in the mid-eighteenth century with irrigation of meadows to provide lush spring and summer pasture and to raise clover. This was an especially important crop, as it matured a month before grass at a time when hay reserves were low, and, moreover, as one farm expert put it in the 1780s, when a cow eats clover it "flushes her to milk." By the 1790s, these new feed crops were common, as was the practice of storing hay for winter feeding. In 1789, Benjamin Rush quoted a German maxim with approval: *Wer gut futtert—gut buttert*: "Whoever feeds well churns much butter." By that time, many farm families in the Philadelphia hinterland had put this principle into practice. In 1804, one traveler reported that farmers near West Chester were reserving fifteen to twenty acres of hay to use for winter fodder.[4]

Once farmers began to increase feed crops, they also paid more attention to its storage. The building of barns occupied the attention of many farmers during the second half of the eighteenth century, and wealthy farmers began to bring feed and then cows into winter shelters. By the time the new federal government took a census of buildings for its first direct tax in 1798, about half the farms had log or frame barns, and a small percentage already had substantial stone barns. Most of the large stone barns that remain today among the finest vernacular architecture in southeastern Pennsylvania and northern Delaware date from the early nineteenth century, when new farming methods resulted in greater amounts of hay stored, investment in more valuable cows, and larger herds.[5]

After the shift in care and feeding of animals, farmers began to experiment with the quality of cows, looking for new breeds that gave better or more milk. The search for a better milk cow preoccupied both gentlemen farmers, who read the latest books and agricultural journals for advice, and country farmers, who gossiped, swapped, and bought their way to better herds. By the 1840s, many dairies in the Philadelphia hinterland had improved Durham shorthorn cows that gave as much as thirty-six quarts a day. In the late eighteenth century, even the most ambitious farmers had been satisfied with eight quarts a day.[6]

The length of lactation had also become an important concern. Medieval cows usually lactated no longer than four months, and probably some of the more poorly tended colonial cows gave milk for only slightly longer periods of time. A well-cared-for English cow produced for an average of eight months in the seventeenth century. By 1845, the *American Agriculturist* was reporting that Durhams could be milked for ten months with a period of only two months dry. The length of lactation increased the duration of the farm woman's work in the farm cycle as well as the intensity. Both cow and butter maker worked longer and harder by the middle of the nineteenth century.[7]

The task of processing greater and greater quantities of milk was a major problem as larger herds giving more milk became common. Rather than one or two cows, the average farm by 1850 probably had about six—the mean number in northern Delaware, where a careful statistical study has recently been completed. Fifty to 75 percent of the farms, depending on the area, each produced four hundred to six hundred pounds of butter a year. Although the job of milking was not difficult, it probably took a minimum of seven minutes per cow for a skilled milker and was usually performed before sunrise so that the cows could move to pasture at daylight. Six cows could take almost an hour to milk. By the late 1840s, dairy books were also advising milkers to take cold water and a cloth with them to clean the udders before milking. Although this was an advance in cleanliness, the procedure added time to the busy milker's schedule. Carrying the six-gallon milk pails back to the house or springhouse could also entail several trips of distances from three hundred to five hundred feet.[8]

Early eighteenth-century farm women probably used cellars for butter making, but by 1800 they were using a springhouse or a milkhouse. Although travelers commented on the presence of springhouses in the Philadelphia hinterland before the Revolution, the houses do not seem to have been common before the late 1760s. From that time on, Philadelphia newspaper advertisements for the sale of farms frequently mention them. The first springhouses were built of logs and were quite small. Most of the larger sturdy stone springhouses still standing in southeastern Pennsylvania and northern Delaware date from the building boom of the early nineteenth century.[9]

The fragmentary direct tax lists of 1798 give some clues as to how many women had these structures at that time. In West Caln, on the boundary of Lancaster County, 53 percent of the houses had what census takers called "milkhouses," small stone or log structures averaging 85 square feet. These small milkhouses seem to have been a type of springhouse used primarily for storage. Similar small milkhouses became common again in the late nineteenth century when creameries began to operate and milk was only stored at the farm rather than being processed into butter there. In areas closer to the Philadelphia butter market, the larger springhouse, a rectangular outbuilding with a distinctive overhanging roof in the front, had already become an important part of women's material culture. The springhouse, according to Henry Glassie, was Continental rather than English in origin. By 1798 more than seven hundred stone springhouses dotted Chester County, with 52 percent of the farms having them; they averaged 127 square feet. In Bradford and Concord townships, 25 percent of the farms had even more generously sized springhouses, averaging 150 square feet. Within these spaces, farm women could process milk close to a plentiful supply of water and in a temperature that allowed easier churning and longer storage time (see Appendix table 17).[10]

The springhouse, as the name implied, was built over a spring, usually into the bank of a hill. Water flowed into a paved sunken trench about two feet wide and three inches deep around a raised center platform. Milk pans were placed in the water to cool, while wooden or stone shelves and benches lining the walls provided space for tubs, bowls, and other processing equipment. Shelves on the outside provided space for drying containers and churns after they had been cleaned. Larger springhouses had two or three compartments, one of which usually housed the spring with the others devoted to processing. The overhanging roof, often seven or eight feet, gave an additional outdoor work space. The springhouse was the woman's domain, where she spent many hours from June to October when cows gave most of their milk and most of the butter was processed.[11]

After 1800, many farm families built even larger springhouses. Peter Hatton built a new 312-square-foot springhouse on his Birmingham farm in 1830. The Pratt farm, in Edgemont, which showed a 120-square-foot springhouse on the 1798 tax list, had a new one of 384 square feet by 1820. Esther Lewis, the Vincent farmer, mentioned building a new 459-square-foot springhouse below her old one in 1837. Although the information on these early nineteenth-century springhouses is fragmentary, it is possible that farm women's work space for butter making may have more than doubled in fifty years.[12]

Straining the milk and pouring it into shallow pans and setting them in a cool place was a process that took more time as milk production increased. Throughout the early nineteenth century, farm women used gravity cream separation, setting milk in shallow pans for forty-eight to sixty hours until the cream had risen to the top. Wooden- and earthenware were most commonly used with tin and glass milk pans appearing in the 1840s. The depth of pans remained constant throughout the period, for they could be no deeper than four to five inches; otherwise the cream would take too long to rise. Cream pans may have increased somewhat in capacity, however, for sources mention pans holding one and one-half gallons and some potters in Chester County sold milk pans as big as fourteen inches in diameter (see fig. 3).[13]

A lead scare after 1785 may have affected the dairy woman's choice of pans. During these years, there was considerable debate in dairy literature about the best materials for cream pans and other utensils. The *Pennsylvania Mercury* of February 4, 1785, raised an alarm about the effects of lead-glazed earthenware on the "country people and the poor everywhere." At that time great quantities of low-cost, lead-glazed pottery was being made and sold locally because stoneware was so expensive. "It is indeed becoming more and more necessary to the calls of the country," continued the writer, "that stone and earthenware should be made and improved and at home." An early farm book by John Bordley, published in Philadelphia in 1801, transmitted the same concern over fifteen years later: that earthenware made in America was glazed with lead and so thin that it could scale off or be worn away by acidulous matter. "It is pure *lead*, and consequently a strong *poison*," he warned. Books published in the 1820s and 1830s continued to warn about the dangers of leaden glazes that could decompose and poison people if the contents of the earthenware became acid. One account suggested scalding and scouring the lead-glazed pans with salt and water daily. Although American dairy women would have agreed with the author of the English *Lady's Country Companion* that "no good dairy-maid . . . would ever

FIGURE 3. **Wooden milk tub for gravity cream separation.** Diameter 15 inches; depth 5 inches, with taller handles. From the Collection of the Mercer Museum of the Bucks County Historical Society.

keep milk in her pans till it became acid," the fact that some methods used slightly sour milk caused great concern. A shift from lead to nonlead glazes for earthenware and the practice of making many dairy-related forms of pottery only in stoneware seem to have followed the scare. At any rate, by the 1830s, potters in Chester County were providing dairy women in the Philadelphia hinterland with safe inexpensive butter equipment for purchase or barter.[14]

Even with safe and larger pans, women still had to handle an increasing number each day. The milk from each milking and sometimes from each cow was kept separately. The cream was skimmed from the surface with a skimmer (a paddle with holes to allow the milk to drain off) and put into cream pots or buckets and then into a barrel, sometimes with a spigot at the bottom to drain off additional milk. Cream was kept in the barrel until enough was available to churn or until it had slightly soured (see figs. 4 and 5). Women churned either once or twice a week in the summer, depending on the size of their churn. Esther Lewis, who had three cows and left a record of her churning for 1830–31, churned every five to seven days, usually about twenty to twenty-five pounds of butter at a time. Larger herds meant more frequent churning. In winter, women usually churned once a week, for there was less milk and cold weather made the cream rise more slowly. Sometimes butter making came to a complete halt in exceptionally cold weather. Lewis wrote in her diary one cold December day: "Cream did not produce butter."[15]

FIGURE 4. **Straining and skimming.** From *The Progress of the Dairy: Descriptive of the Making of Butter and Cheese for the Information of Youth* (New York: Samuel Wood, 1819). The Sinclair Hamilton Collection of American Illustrated Books, Princeton University Library.

Increased work space allowed more processing to take place, but the duration of the processing probably increased in proportion. Concern focused on churning—to lessen the labor and to increase the amount of butter produced. The time to churn decreased with both the barrel and the box churns and particularly with temperature control, which appeared in dairy literature as an important factor in the 1820s. While the atmospheric churn reduced churning time to less than ten minutes in the late 1840s, it did not produce quality butter; apparently the time required for really good butter ranged from forty-five minutes to an hour and a half. Colder temperatures could delay the process for hours, even with an improved churn. The importance of even, regular, and consistent strokes with dasher or crank also made this a job that needed undivided attention.[16]

FIGURE 5. **Sour cream tub for collecting cream for churning.** From the Collection of the Mercer Museum of the Bucks County Historical Society.

Churns, regardless of their mechanical details, usually have two parts: a container for holding cream and a mechanism for agitating the cream to separate the fat particles in milk from the watery fluid in which the fat is distributed. Technically, churning breaks the casein shells in which the fat is enclosed, thus separating the butterfat from the watery liquid and gathering the fat particles into a mass by agitation. As one modern writer put it, churning converts the emulsion from water in oil to oil in water and coalesces the fat globules. Colloquially, the process was said "to bring the butter." When the separation was complete, the butter had "come." Few studies of seventeenth-century inventories in the Pennsylvania hinterland mention churns, but by the eighteenth century, they frequently show up in the inventories of various classes. The spread of churns seems to have accompanied improved pasturage and care of milk cows in the eighteenth century and the rising demand for butter.[17]

The first colonial churns were probably plunger or dasher churns, a type that dates back to fifteenth-century England. The churn was a great saving in time and a great convenience compared to butter making with no specialized equipment. The plunger churn, with variations, was in use in America for over two hundred years. Yet, because the history of this specialized tool used by generations of women has not yet been studied in detail, it is still difficult to date precisely the different versions. The plunge churn consisted of some type of tapered receptacle for the cream and a plunger with various devices at its foot to agitate the cream. Coopers made these early churns by hand, usually of staves, which were held together with willow wythes fastened by cord or nail, and rather simple plungers, consisting of a circular disk or two crossed wood pieces either plain or with holes in them. Some early churns were only eighteen inches high, held a gallon of cream, produced one or two pounds of butter at a time, and took three hours to operate. Although women used these simple churns for individual household production throughout the nineteenth century, those who produced for the market needed larger churns so they could produce more butter at one time. They also needed churns that would produce butter in a shorter time with less effort.[18]

Even simple churns varied greatly. Tin churns were used after the spread of tinware in the 1820s. Potters regularly produced stoneware crocks with lids that had holes for the insertion of a plunger. Wooden churns followed regional patterns and changes in the cooper's trade. Some had willow wythes or wooden hoops, and later many had machine-made iron hoops. By the mid-nineteenth century, factories mass-produced churns by machine in large quantities. From the 1890s, farm catalogs advertised wooden dasher churns in a great variety of sizes. In 1899, for example, James S. Barron and Company sold cedar dash churns from twenty-six inches to sixty inches high that held from three to ten gallons of cream and larger oak churns that held from seven gallons to two barrels (over sixty gallons) of cream.[19]

The market demand soon led to the introduction of new types of churns, but the plunger type remained in common use in small dairies because of its cheapness and its convenience for cleaning and getting out butter. The churner had to maintain an even stroke with the plunge churn, and when the cream became thick, a vacuum formed at the bottom when raising the dasher. The operator then not only had additional labor in raising the plunger but also usually needed someone to hold the churn down. The dasher churn was not suitable for making large quantities of butter.[20]

The pump churn, an early variation on the plunge churn, used the mechanical advantage of a lever to partially solve the vacuum problem. The pump churn could also easily be put in water to control the temperature, but it had the same disadvantage of the plunge churn in that a regular motion had to be maintained. An early dairy adviser, James Cutbush, warned users in 1814 to "by no means admit any person to assist them unless from absolute necessity; for, if the churning be irregularly performed, the butter in winter will *go back*; and, if the agitation be more quick and violent in summer, it will cause the butter to ferment, and thus to acquire a very disagreeable flavour." Whether this was true or not, it encouraged the practice of one woman assuming responsibility for the churning process. How then could they continue to process butter by themselves and produce more for the market?[21]

FIGURE 6. **Churning.** From *The Progress of the Dairy: Descriptive of the Making of Butter and Cheese for the Information of Youth* (New York: Samuel Wood, 1819). The Sinclair Hamilton Collection of American Illustrated Books, Princeton University Library.

One response to the market demand for butter was the invention of the barrel churn. Introduced in the early eighteenth century in England, barrel churns show up in Chester County inventories of the 1750s. For years dairy books listed dasher and barrel churns as the main types in use (see fig. 6). *The Progress of the Dairy*, published in 1819, described the barrel churn simply as "like a large barrel, fixed on a stand; with a handle to turn it round, and a small square door into which the cream is poured." The barrel churn allowed women to churn greater quantities of cream at one time. It could also be adapted for use with sheep or dog power, although there is little evidence to show how early women began using animal power for this purpose. The *Farmer's Cabinet* in 1837 noted a barrel churn could be used to produce thirty-six to thirty-eight pounds of butter a week. From a mechanical standpoint, it seemed to be an improvement because the operator could keep all the cream in constant agitation much as a modern clothes dryer tumbles clothing. With a dasher churn, by contrast, the operator could agitate only the area around the plunger and had to stop the motion, however briefly, at the top and bottom of each stroke. The barrel churn remained in use through most of the nineteenth century. One variation, developed and used widely in the Philadelphia hinterland, became known as the Philadelphia churn, which could churn seventy gallons of cream in one and a half hours.[22]

Barrel churns did not eliminate the skill needed to operate them. All churns had to be kept very clean, and the larger barrel churn was more cumbersome and had more internal parts to wash. The operator still had to exercise care in operation, turning in the same direction and at a constant and regular speed. One commentator in the 1847 *The Book of the Farm* warned: "The rate of motion in churning butter is of some importance for, when performed too slowly, a longer time will be spent on churning than is necessary, and the butter will be strong-tasted; and, on the other hand, when the motion is too rapid, the butter will be soft and frothy, when the churning is said to have *burst*." He still recommended one and a half hours as the most satisfactory churning time, even with a barrel churn, so that the butter would be neither too soft nor too strong-tasting.[23]

By the 1830s, women were increasingly unhappy over the difficulty of cleaning the barrel churn. Moreover, interest was growing in the concepts of a movable body and the elimination of the dasher. One enterprising inventor had patented a rocker churn in 1803, and other variations using a plain box as churn soon followed. The box churn, in principle, was simply a square box set on some type of stand that could be either cranked or rocked (see fig. 7). Contemporaries considered it a major breakthrough in churn technology because it used concussion in addition to motion to bring the butter—the corners and right angles compounding the motion. It is difficult to tell just how popular the box churn became, but it does frequently appear in museum collections. One dairy expert claimed that a box churn could be built "by anyone who can handle a saw and plain." But this expert was certainly overly sanguine, for if made by someone who was not skilled in carpentry, the box churn could leak and crack as did some barrel churns after long use. Still, the box churn, and especially the rocker version, may have been popular before the atmospheric churn was introduced at the end of the 1840s.[24]

Churn technology preoccupied many people in the early nineteenth century. In fact, this period could be characterized as one of quest for the "great churn." From 1802 to 1849 the Office of Patents issued 244 patents on machinery related to butter making, 86 percent of them for churns. The search seemed to intensify in 1807–11, 1827–38, and es-

FIGURE 7. **Box churn.** Height 35½ inches. From the collection of the Mercer Museum of the Bucks County Historical Society.

pecially 1848–50, when over a quarter of the patents were issued. Although modest compared to the boom of 1850–73 when 1,360 butter-related inventions were patented, the early nineteenth-century quest for better machinery is impressive.[25] There is no way to tell from the patent listings themselves whether these new inventions were of any value. But Thomas P. Jones, editor of the *Franklin Journal*, reviewed 92 patents, almost half of all the churns patented. He was not impressed. He published 35 percent without comment, 61 percent with unfavorable comments, and only 4 percent with some positive response.

Jones found little novelty in most churn patents. He occasionally commented favorably on dasher improvements, but most patents he damned as offering nothing new. One can almost hear the editor sigh as he wrote in 1832: "Further we need not say, as the individual parts possess no more novelty than does the general construction; and those who have read our lucubrations, for a short period only, will be able to pronounce at once upon this point." Or later that year: "We have had a truce with churns for some time past, and regret that in again introducing one to our readers, we are compelled to say that the character of novelty which is claimed for it, cannot be sustained." And with exasperation in 1833: "The whole churn is not better than a thousand others." And in 1836: "The particular arrangement of the floats, and the advantages which the patentee thinks will be derived from them we leave to be recorded by some future historians."[26]

But just what can the "future historian" now make of this wit and sarcasm lavished on the stumbling inventors of these early churns? There were no patent lawyers or patent research to allow the isolated inventors access to information about previous patents. The inventor simply paid thirty dollars and deposited the patent model and claim at the Patent Office. In this early period, the office merely registered all submitted ideas, rather than just new ones. Moreover, a high proportion of patentees came from rural New York and New England, areas that produced large quantities of butter but were less known for the quality of their butter than was Pennsylvania. It was, however, precisely in these areas that the earliest factory-produced butter emerged.[27]

Churn reviews can be read in another way. Patentees were searching for ways to substitute mechanical power for woman power in churning. The patents reveal a preoccupation with efforts to churn larger amounts of cream faster, to form the butter into a more solid mass for easier working, to reduce the skill needed in churning by having the machinery respond to the consistency of the butter as it formed, to introduce less expensive construction (especially in substituting square construction for barrel construction, which required a cooper's skills), to control temperature, to use feet instead of hands, and to create a machine for both washing and churning. These efforts reflected a demand for improved machinery, even if they did not prove successful.[28]

None of this early patent activity involved women. Of the thirty-five women who applied for patents between 1809 and 1850, not one designed equipment for this predominantly female work. A woman was more likely to patent a sheet-iron shovel, a submarine telescope, or even locomotive wheels than machinery to make butter making more efficient. The first woman to patent a butter-related invention was Lettie A. Smith, a thirty-six-year-old, single Quaker farm woman from Bucks County who applied for a patent on an improved butter worker in 1850. Smith went on to medical school soon after, and only two or three other women patented churns in the great patent rush of 1850 to 1873 (see fig. 8).[29]

This lack of patent activity by women does not mean they were not interested in more efficient methods. Indeed, they may have participated in the experimentation that preceded the patent action. But most women were not part of the new public networks of male mechanics and would have been more likely to have communicated any improvements by word of mouth and demonstration rather than through patent activities. Women

FIGURE 8. **Model of butter worker and scales patented by Lettie Smith in 1853.** Box 30½ inches square. From the collection of the Mercer Museum of the Bucks County Historical Society.

did compete for butter prizes in the fairs set up by the new male agricultural societies of the 1830s, but there were no churning competitions where they matched themselves in the process rather than the product. In fact, the major churning matches seem to have occurred in the late 1840s and to have involved the newly discovered atmospheric churn.

This churn was based on the principle that air rather than concussion induces fat particles to combine. The *Scientific American* reported that an atmospheric churn produced butter from cream in ten minutes at a churning match held in 1848. But at this trial the mechanics seemed intent on time rather than quality, and the editor disclaimed the benefits of rapid churning, noting "we should prefer that the churning, for a quantity of ten to twenty pounds of butter or more, should be prolonged to thirty minutes, at least." The introduction of the fast atmospheric churn, named because it forced air through the cream by means of a hollow upright shaft, came near the end of nearly a century of continuous changes and improvements in churns and led the way to eventual mechanization of the process.[30]

Churning was only one part of the process of making butter. Working the butter was acknowledged by most commentators to be even more important than the actual churning. Like churning, butter-working techniques changed over the years in response to the market. English immigrant women brought a great variety of regional butter-making skills with them to the New World. By the seventeenth century, butter had become a market product in England where large quantities regularly were shipped to London. Women had to learn not only how to process larger quantities of butter more quickly but also how to successfully preserve it for longer periods of time and to improve the quality so their butter could bring premium prices in the marketplace.

Butter working was the process of kneading out the buttermilk, working in salt or a preservative, and compacting the butter into a solid mass. The importance of working is evident in the appearance of improved butter-working equipment and in the advice that farm books and journals offered to the inexperienced people who were moving into farming in the early nineteenth century. Traditionally, after learning from mother or employer, each woman developed her own techniques. Temperature was important in this process, too. Skilled Pennsylvania butter women might keep a cold marble slab or a dish of ice water into which they could dip their hands to keep them cool. There was a folk saying: A woman with cool hands makes good butter. Or as one authority warned: "A woman who has hot, clammy hands should never become a dairymaid."[31]

As butter became an important item of household produce, women learned not to handle the butter with their hands at all. As early as 1803, Thomas's *Almanak* damned "the beating up of the butter by the hand" as "an indelicate and barbarous practice." More than manners were involved, however, for butter worked with the hands was more difficult to shape than butter worked with some sort of tool. Women used at first two flat boards, then a flat butter paddle, and then more elaborate butter workers so that the liquid could be pressed out without the problem of keeping the butter cool during working. Improved butter workers, like the one Lettie Smith patented in 1853, were important in shortening the amount of time needed to work a large quantity of butter, for the process, as one commentator remarked in 1826, took "dexterity as well as strength." The large butter worker enabled farm women to substitute a tool for a special skill. The small butter worker, however, remained a symbol of both women's skill and the handing on of that skill from mother to daughter. The Mercer Museum in Doylestown, Pennsylvania, contains in its acquisition files two notes that represent this tradition. One wooden paddle marked 1833 is accompanied by a note that a grandmother received it as part of her wedding gifts. A second butter worker carries the notation that it was passed down from mother to daughter until the granddaughter deposited it in the museum in 1907 (see figs. 9 and 10).[32]

Recipes for preserving butter appeared in the earliest dairy books of the nineteenth century and persisted through the 1850s. After 1800, a saltpeter-salt-sugar recipe quickly replaced plain salt as a preservative. The new method allowed butter to be kept for up to three years, whereas the earlier brine method seems not to have kept butter fresh much more than a year. A few recipes for pickling butter in brine continued to circulate after the 1820s, but most of them disappeared from the literature and the saltpeter-salt-sugar method predominated. How much fresh and how much preserved butter was produced during those years is still to be determined.[33]

Whether fresh or preserved, the butter had to be packaged and prepared for transport to market. Surprisingly, the butter print, an artifact almost as ubiquitous as the churn, is seldom mentioned in early accounts or inventories. Thus more research needs to be done to classify prints accurately as to region and date. One 1819 picture shows a woman clamping a roll of butter down on a print (see fig. 11). A Pennsylvania newspaper advertisement of 1828 offered "100 dozen Butter Prints" for sale, an indication that the mass distribution of prints had begun by that time. An 1831 article mentioned that any butter to be marketed soon should be made into half-pound or one-pound cakes and marked. Martha Ogle Forman, who was sending butter to the Baltimore market in 1830, recorded receiving two butter prints from Baltimore, and Esther Lewis talked about printing some of her butter for the Philadelphia market in 1831. It seems likely that the increased butter making of the 1830s was accompanied by an increase in the manufacture, sale, and use of butter prints.

When shipping only small quantities of butter, like those Lewis sent to Philadelphia, women put their printed cakes into the refrigerated pails or boxes that appeared soon after 1800. Called the Philadelphia butter pail by the 1850s, these pails and similar containers had compartments for ice so that butter would reach the market fresh and firm during the warm summer months (see fig. 12).[34] Both Forman and Lewis also record in their farm accounts that crocks or pots were used for sending butter to market. Some of these crocks were quite large and could hold twenty to twenty-five pounds of butter. Other containers used included wooden tubs that held ten to eighty pounds and firkins that would take as much as a hundred pounds.

The prescriptive advice on how to pack the butter was detailed. Although occasionally the use of glazed pots was still recommended in spite of the lead danger, most experts advised that wooden firkins be used.

FIGURE 9. **Butter-working table.** The tray is made of pine, and the rest of oak. Grooved to carry off whey. Rear 25¾ inches high; tray nearly 23¼ inches; front 13⅛ inches; length 39 inches. From the collection of the Mercer Museum of the Bucks County Historical Society.

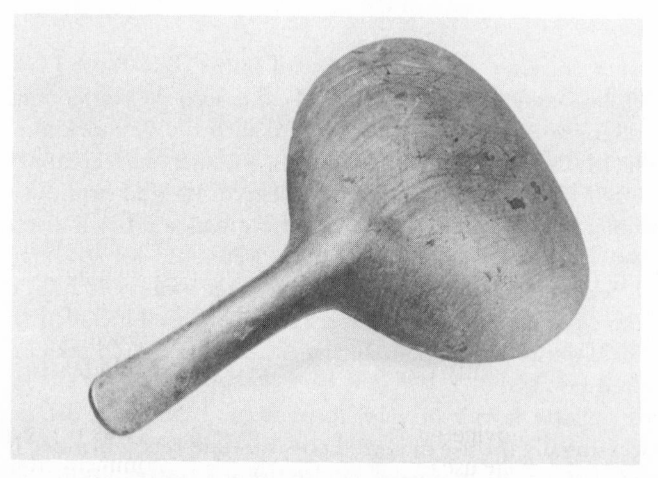

FIGURE 10. **Butter paddle.** Length 11½ inches; width 6¾ inches. From the collection of the Mercer Museum of the Bucks County Historical Society.

FIGURE 11. **Weighing and printing butter.** From *The Progress of the Dairy: Descriptive of the Making of Butter and Cheese for the Information of Youth* (New York: Samuel Wood, 1819). The Sinclair Hamilton Collection of American Illustrated Books, Princeton University Library.

These containers were to be well seasoned by scalding and rubbing with salt and their cracks sealed with melted butter. Then, after packing, either a cloth dipped in butter or a layer of melted butter was placed over the top to seal out air. The butter was then ready for market.

Brissot de Warville described the Philadelphia market in 1788 with women bringing in their produce from the country in "handsomely woven baskets." Wagons and horses lined up in neighboring streets in order of arrival, and everyone bought and sold quietly without shouting. Police clerks kept an eye on the produce. If they suspected a pound of butter was short, he noted, they weighed it, and if indeed it was short, they seized it and gave it to the hospitals. A male relative or tenant farmer might take several women's butter with him to Philadelphia. Farm women who lived on rivers could load their crocks and firkins on the docks to

FIGURE 12. **Portable butter box for carrying printed butter to market.** From the Collection of the Mercer Museum of the Bucks County Historical Society.

be sold by riverboat captains in Baltimore or Wilmington. But most of the time, the women carried their product to market and offered the butter for sale out of pails and boxes they had carried on horseback or from the ends of wagons backed up to the curbs along the streets of town.[35]

Family members seldom returned from market with cash; rather they would bring back commodities. When Esther Lewis sent butter and other farm products to Philadelphia in the 1830s, marketers brought back salted fish, sugar, tobacco, coffee, pepper, textiles, sewing supplies, and books. Lewis also traded her butter locally with neighbors for work like shoemaking and repairs, with rural potters for earthenware, and with country merchants for cloth, tinware, and other commodities. And she probably traded it to the "pedler women" who, she noted, visited her farm in 1830. Butter was not the only commodity sold and bartered; it was only one item in a complex web of commodity exchange that existed within the community of kin, neighbors, church members, and crossroads merchants. It was, however, an increasingly common item of trade during the early nineteenth century.[36]

During the century from 1750 to 1850, then, the dairy work of both women and men changed significantly. From a peripheral part of farm work, dairying moved to a central place on many of the farms of southeastern Pennsylvania and northern Delaware. By 1850, 4,760 farms in the counties of Delaware, Chester, and New Castle were producing 4 million pounds of butter. Rural women, for better or worse, were tied directly to the emerging commercial capitalism of the late eighteenth and early nineteenth centuries. Butter making was thus an important process by which rural women extended the old concept of the productive farm economy into the new industrial era.

CHAPTER 7

Rural Domestic Economy

The crucial role of women in reproducing the farm family, providing labor, developing the economy, and refining the technology of butter making did not mean that these important functions would necessarily be reflected in the literature of the time. Although literature is the mirror of a people, it reflects only what those who write it wish to see. Literature extends the experience of life; thus it is a chief way of measuring the significance of one's life and beliefs or ideology. Embedded in literature are the wishes and needs of the group that produces it, so that it may provide only a distorted and indistinct mirror of the lives of those who do not control it. Nonetheless, the literature of a dominant group can still reveal changes in the status of other groups. Thus the writings of men can tell us something of the role of women in the rural economy.[1]

Historian Mary Beth Norton has pointed out that there was little discussion of women's public roles until the late eighteenth century. Laurel Ulrich has shown that Indian captivity provided one way in which New England women might become public heroines. Quaker women had a particularly important public role in their religion. Most women, however, were expected by their communities to remain within the private household sphere. Rural women's skills and roles were transmitted primarily through the oral tradition in private rather than in public discourse. The secrets of housewifery were indeed secrets. Indentured women paid with their labor to learn from mistresses; daughters learned from mothers. Existing along side of this functional oral tradition was another one that included story, fable, proverb, humorous anecdote. This second tradition offered guidelines for social relations by providing role models, outlets for repressed desires, arenas for revenge, and prescriptions for worldly success.[2]

These eighteenth-century secular concerns were seldom committed to public discourse in the form of published literature. The Bible offered contradictory prescriptions and role models that were adapted to the appropriate time and space through the mediation of the oral tradition. Ministers interpreted the Bible through religious didactic literature, but the many daily concerns of the rural population, the adjustments of skills and decision making, were mediated by the oral tradition, the folklore of the people. In the late eighteenth and early nineteenth centuries, much of this folklore was written down as the middle class developed its written literature and fashioned a culture to meet its needs.[3]

Little folklore remains from the Quaker culture of southeastern Pennsylvania, unlike the Pennsylvania Germans, the Irish, and the Africans, who left distinct folk narratives. Richard Bauman argues that because seventeenth-century Quakers condemned foolish talking, vain jesting, profane babbling, and fabulous stories, a wide range of folk narrative was not acceptable to them and was suppressed or ignored in the development of their culture.[4]

Although much folklore of the Quaker culture area may have been lost, remnants of it survived in the first mass popular culture developed in early eighteenth-century almanacs. Embedded in the proverbs that Benjamin Franklin made famous was advice on how to conduct one's material life and social relations in order to be successful. Franklin published his *Poor Richard's Almanack* from 1732 to 1757, which established the genre. About ten thousand *Poor Richard* almanacs were printed annually in the period from 1751 to 1765, and over three thousand almanacs were published in English and German in Pennsylvania alone between 1750 and 1850.[5]

Almanacs contained three basic elements: a calendar, an astronomical computation of passage of time, and prognostication of weather. To fill up the rest of the pages, editors used stories, advice on farming practices, and short bits of folk wisdom. Nineteenth-century almanacs also often carried lists of officials and information on other states. In Philadelphia they sometimes carried the dates of the Quaker quarterly and yearly meetings. A number of women published almanacs before 1750, of whom the best known were C. Zenger, Cornelia Bradford, and Ann Franklin. Lydia R. Bailey printed several almanacs in the early nineteenth century. Almost all of them, however, were compiled by men. Rural people purchased almanacs because they were inexpensive and readily available at country stores or at the Philadelphia market where they sold produce. Judging from the names written on the outside of almanacs, we know that women owned them individually as well as having access to those brought home in the farm wagons as part of the family provisions. Most of the 167 almanacs examined for this study were published in or near Philadelphia and were intended for use by rural people in the surrounding hinterland. They comprise about 5 percent of all the English almanacs produced in Pennsylvania between 1770 and 1850. Almost half of them were published between 1820 and 1840.[6]

Publishers of almanacs found their material in a variety of places. Some, like Franklin, wrote their own. Others selected stories and anecdotes from a store of previously published literature that was thought to appeal to rural folk. They ranged from what appear to be traditional English and French folktales set in earlier agrarian times to stories borrowed from recent publications. Most of the stories were anonymous; only a few were attributed to an author. This popular literature was transmitted much as oral tradition was, with the transmitter exercising the prerogative not only to select but to change. There was almost no reference to religion or to spirituality, no advice to either men or women to be pious, nor was there any comment on politics. Political literature developed as a separate genre during the eighteenth century, in newspapers, broadsides, and pamphlets. Rural males provided an audience for the political literature of the Revolution and later for the new party newspapers. Nor did almanacs include advice on motherhood. The closest any came to this subject was to offer a cure for sore breasts. Thus literature aimed solely at either males or females was excluded. The stories instead concentrated on domestic relationships, often dealing with gender conflict. Some counseled female submissiveness, but others showed women displaying a considerable amount of assertiveness.[7]

The contradictory attitudes toward women in almanacs may, in part, be explained by the fact that almanac literature reflected a literary tradition that was in transition from oral to written form. Folklorists have noted the presence of assertive heroines, particularly in the Anglo-American folktale and to a lesser degree in German folktales. They have also noted a tendency for assertive folk heroines to be transformed as the oral tradition was written down, a process usually performed by educated males. As the literature was prepared for mass circulation, heroines were further transformed. The classic example of this is the German Grimm brothers' folktales, which were first collected in 1812 in dialect, many coming from women servants. Some tales were stylized and transcribed into High German and published. Of the original 210 stories, 40 contained heroines of assorted types. Only a handful of these heroines, the most docile, normally appeared in English translations. In the twentieth century, as Kay Stone has noted, Walt Disney in turn selected the prettiest and most passive of these heroines, the ones who were patient, obedient, industrious, and quiet. Aggressive, ugly females appeared, usually in the role of witches or stepsisters.[8]

Scholars find Anglo-American heroines generally more assertive than those of the Grimms, and it is this tradition that the Pennsylvania almanacs seem to reflect. The didactic advice carries hints of female assertiveness along with general advice to be submissive. Gender differences, according to these almanacs, could be reconciled by developing a type of domesticity in which both men and women worked together for the common cause of familial success, not just in economic but in social terms as well.

Gender conflict outside of marriage is sometimes presented in almanacs. Abduction was the theme of one story titled "The Lady and the Robbers." Here a baroness "scarcely 20" promised to go with a robber to save her life, clearly implying the choice of dishonor over death. Whereas a married woman was perhaps allowed more freedom in securing her life than was a maiden, this clever woman managed to imprison the robbers in the cellar by luring them there and then pushing the guard down after, an act described in this way: "She made a spring at the wretch who as little expected the dissolution of the world as such an attack. A single push with all her strength tumbled him down stairs from the top to the bottom. In a twinkling she closed the trap door—bolted it—and thus the whole company were secured in the cellar. She then roused the neighboring farmers and servants and had them kill the robbers."[9]

Seduction and abandonment was the subject of several stories. Remorse by males after killing young women to conceal their pregnancy was a common theme of traditional ballads, which were usually portrayed as being composed as the men faced the gallows. In these Pennsylvania almanacs, seduction and abandonment led to dire consequences, although not to the gallows. The *Columbian Almanac* of 1801 carried a story entitled "Maria, a fragment—founded on fact," a sentimental tale of woe by a dying girl "corrupted and destroyed," whose seducer, seized with remorse, plunged his sword into his breast and died at her side. Maria, although complaining, has no part in her lover's death. In "Miss Bailey" and "Kitty Maggs and Jolter Giles," both published in 1812, the old theme of seduced maidens is depicted in ballad style. "Miss Bailey" began: "A Captain bold in Halifax, / Who lived in country courters, / Seduc'd a maid that hang'd herself / One morning in her garters." The second began: "Kitty Maggs was a servant to Farmer Styles, / and a buxom wench was she, / and her true lover was Jolter Giles, / ploughman so bold was he." Both women are betrayed and die, but each returns from the grave to wreak revenge. Miss Bailey returns and takes the captain's leather breeches with a one-pound note in them. Kitty Maggs chokes Giles by forcing pudding down his throat. These are traditional revenge stories of wronged women; nevertheless they portray women as aggressive ghosts. One happy ending for a seduced maiden perhaps shows the changing attitude toward illegitimate birth by 1799. Here a young woman is classically seduced and deserted. She plans to murder her child and kill herself, but instead only abandons the infant. The baby is found alive and preserved by a stranger, and the mother is reunited with her lost child years later.[10]

This punishment for males and females engaging in sexual activity outside of marriage reinforced the overwhelming advice that young people marry. Although early almanacs from the 1780s and 1790s included more negative comments about women than those published later, the most negative were reserved for bachelors and old maids who refused to marry. Marriage was portrayed as the best job for both men and women. The exceptions were glaring. *Poor Richard* had a man choose the gallows over marrying: "Lo! here's the Bride, and there's the Tree, / Take which of these best likest thee, / The Bargain's bad on either Part; / The Woman's worst;—drive on the Cart." *The New-Jersey and Pennsylvania Almanac* of 1814 likewise had a Viennese woman choose death over an ugly husband. Such attitudes were the exception.[11]

Even before marriage, however, the question of female subordination surfaced. Biorin's almanac of 1813 denounced "petticoated philosophers, blustering heroines, or virago queens," and in 1815 he listed mildness, contentment, innocence, and modesty as requisites for a lady. An 1829 almanac reprinted Madame de Maintenon's "Advice to a Bride," cautioning her not to hope for too much. "Men are tyrants, who would be free themselves, and have us confined. You need not be at pains to examine whether their rights be well founded; it is enough if they are established." She then went on to advise that affections could not be gained by "complaints, reproaches, or sullen behavior." Another story advised the prospective bride to exhibit "Female Spirit." Before marriage, the groom said he would sleep alone, eat alone, and find fault when there was no occasion. The bride replied that if he slept alone, she would not, if he ate alone, she would eat first, but that he would never lack an occasion for finding fault. Perhaps the most graphic metaphor for the potential problems of a prospective groom who claimed too much domination was "Peter Going to the Wedding." Peter puts his wife behind him on an ass, "For, says Peter, the Woman, she should / follow, not lead, through life." The ass bucked them both off.[12]

Once married, of course, the potential for conflict increased. Almanacs offered both males and females advice, often contradictory, on how to survive marriage happily. *Poor Richard Improved* for 1793 offered civility as a hint to the married, not specifying either sex. *Father Abraham's Almanac* of 1794 advised men to treat their wives as "reasonable creatures" to obtain personal esteem and affection. In 1781, *Father Abraham's Almanack* passed on the advice of Alfonso, king of Aragon, "On women," condescending in tone and concluding that to make a happy marriage the husband should be deaf and the wife blind. One 1802 almanac showed through an anecdote that a wife who tried only to please a husband was not likely to convince him of his errors; she had to show him to be a fool. *Poor Richard* in 1811 warned that, like princes, women who tyrannized would be abused. The *New St. Tamany Almanac* of 1818 announced to women, "thou art made man's reasonable companion, not the slave of his passion." But the 1837 *Columbian Almanac* asserted, "A woman's love is like the plant which shows its strength the more it is trodden on."[13]

Several almanacs hint that women's refusal to have children might be one cause of marital discord. *Father Abraham's Almanac* of 1794 referred favorably to the Biblical Rachel for wanting children, but complained about "hundreds of her descendants, who cry out, give me no more children, or else I die." In an 1823 almanac, the story "Philander and Eloisa" tells of a wife who left her husband to live in poverty rather than have children. The estranged wife kept "her chastity inviolate" while playing "the coquette" for four years. When about to take a lover, she was saved by a violent fever and subsequently reunited with her husband.[14]

There were examples, also, of how women dealt with husbands who physically abused them. Two wives illustrated the possibilities for curbing spousal abuse. The first was female solidarity. In "How to Tame a Husband," neighbor women informed the third wife how her new husband had physically abused his first two wives. When he was about to beat his new wife, she stopped him with the argument that the right to physically abuse her was not in their marriage contract, and they then agreed to separate. While dividing their possessions, however, she pushed him into a trunk and locked it, and then invited the neighbors in to have tea on the trunk. After hearing their collective discussion of his brutality, the husband begged the women's pardon. "The ladies were so good as to forgive him, and let him out of the trunk," the story concluded.[15]

A second wife used verbal assertiveness and separation. This account told of a Missouri husband who married his wife for the eight black slaves in her dowry. The wife was likewise unwilling to take physical punishment. The husband expected to beat his new wife with a hickory switch occasionally, but found that she was "willing to tally word for word and blow for blow." One day when he ordered her to make coffee, she told him, "Go to hell!" When he threatened to beat her, she retorted: "I'm determined to bear no more of your ill usage. Instead of using mild and conciliating language which a husband ought to use, you always endeavor to beat me into measures—touch me with that whip, I will leave your house." The next morning she left.[16]

These stories portray a model of an assertive woman within a potentially repressive marriage structure that assumed submission but not abuse as the norm. But the message of subordination was less clear than the admonition to be a frugal housewife. The frugal housewife model of almanacs was one of a shared venture in which the woman was to be judged by her attention to household economy.

Almanacs offered several types of advice on domestic economy to couples. The most important was choosing a wife who subscribed to the principles of economy. In 1814 *Poor Robins Almanac* published the parable of "The Choice of a Wife by Eating Cheese." One woman ate the rind, another threw it away, but the third who scraped her cheese was the best choice.[17]

Domestic economy usually began with marriage, however. The *New Jersey Almanack* of 1785 published an anecdote made famous by later historians who delighted in quoting it as evidence of a subsistence economy. The story was narrated by a farmer whose hard-earned fortune was dissipated by the purchase of consumer items and increased consumption of farm products. "My butter, which used to go to market, and brought money, is now expended at the teatable." he complained. This farmer was determined to return to the old subsistence way of life he said was practiced in the 1760s. The *Poor Richard Improved Almanac* of 1793 mocked "The Fair Economist," Flavia, who cheated herself through false economy. Mocking Flavia, the poem reads: "That set of china was the cheapest thing! . . . This cambric . . . Was *such* a bargain." The poem ends with the warning: "But Flavia, stop in time; too late, I fear / You'll find these bargains cost you plaguy [vexation] dear." The *Pennsylvania Almanac* of 1830 lauded "The Happy Match," a household where the husband brought in the money, and the wife kept it from going out foolishly. Choosing each other for love first, and then for being "sensible, economical and industrious," this couple's interests were identical. The wife was expected to make plainness and frugality, neatness and usefulness, her goals, according to this story, which concluded that "while few people lived more comfortably, none lived more economically," and the husband and wife "mutually give each other the credit for doing all this."[18]

This advice continued to appear throughout the early nineteenth century. *Uncle Sam's Large Almanack* carried "Brother Jonathan's Wife's Advice to her Daughter," in which she talked of educating her daughter to habits of "industry, frugality, economy, and neatness." Still the message here was also that the man provided and the woman managed "her own business," the house. Or as the *Agricultural Almanac* of 1840 put it, the art of economy in domestic life was "to turn every thing to account" and "to make the most of what you have." Rapp's *House-keeper's and Farmers' Temperance Almanack* of 1844, borrowing from the popular New England tract by Lydia M. Child, summed up the idea of domestic economy in this way: "The true economy of housekeeping is simply the art of gathering up all the fragments, so that nothing be lost—fragments of time, as well as materials."[19]

A note on the changing definition of *domestic* in the eighteenth century is necessary here in understanding the change in emphasis implied in these prescriptions. When used as a subject in the sixteenth century, the word *domestic* had meant a servant or retainer. By the mid-nineteenth century, it was used to refer to a woman who did chores, primarily inside the house. At the same time, the adjective *domestic* also changed meaning. *Domestic* could refer to tame animals, one's own country, being attached to home or home life, or belonging to the house. It did not, in the seventeenth century, refer to women's work or affairs but rather to activities around the home. The terms *rural economy* and *domestic economy* were both being used by 1820, but the latter referred to management of the household as a part of the overall farm economy and as a crucial part of the overall success of the farm. *Rural domestic economy* did not mean domesticity, in the mid-nineteenth century sense of a quality or state of being domestic. It was, rather, a place where certain tasks were performed, and those were not narrowly but broadly conceived. Nor did it concern only women. As males became more concerned with the market and commercial success, however, women were expected to take a more important role in implementing domestic economy.[20]

This emphasis on women's role in saving and managing pervaded the new literature of agricultural improvement that began to supplement almanacs after the 1780s. These books and journals were obviously aimed at a more highly educated reader than were the almanacs. They were longer, more systematic, and had fewer anecdotes; but they carried essentially the same economic message as did the almanacs—that domestic economy was important to both men and women. This literature was different from almanacs in both distribution and use. Beginning in the late eighteenth century, the number of these books increased as writers tried to apply new concepts of natural philosophy to old agricultural traditions. By the late eighteenth century, English commentators had come to believe that the productivity of the rural poor could be a factor in national prosperity and, hence, that the farm activities of cottagers were worth commenting upon. A simultaneous country life movement occurred in which people without farming experience wanted to supplement urban incomes with produce and sales from country estates. Much of the early nineteenth-century rural domestic economy literature comes out of this tradition. The authors were usually men who conceived of themselves as operating in the experimental philosophy tradition, questioned rural people about their farming practices, and then translated this oral tradition into written form. Like the almanacs, these more expensive books and journals were published in cities, but they reached only the wealthier farmers. They also reflected a changed attitude toward women's role in the rural economy.[21]

Sometimes the writers of this agricultural development literature were explicit in their criticism of the old traditions by which mother trained daughter, particularly in dairying. J. Twambly, for example, in a 1784 treatise *Dairying Exemplified* criticized the traditional mother-daughter system of learning for never "calling in the assistance of either Philosophy, by which they might learn the different qualities, and effect of materials they use, or knowledge, how to apply them in a Physical, or Practical manner." John Bordley, writing his *Essays and Notes* in 1799, also felt that outsiders with "unshackled minds" and "tested principles" could bring great change to agricultural methods. Bordley emphasized the important work of female cottagers in rural Pennsylvania who could produce food in the cottage garden and manufacture flax as well as tend the cows while the men tended to the fields. Twelve hired men, women, and children, he argued, could do the work of thirty-five slaves, and more cheaply. Bordley marshaled figures on wages and living costs as evidence. The productive work of hired women was essential in all his estimates.[22]

Books such as Bordley's combined ideological justification of the importance of an efficient rural working class with hints for the middle-class farmer. Recipes for breads and concoctions to be used against vermin were a part of the detailed prescriptions for a well-managed rural domestic economy. Attentive personal application together with regular accounts, proper information, and accumulation of investment capital had become the formula for success by the early nineteenth century.

Even the physical health of rural women was now cause for public concern. Thomas Cooper's *A Treatise on Domestic Medicine* published in 1824, exemplified the importance of domestic economy and health. Cooper included diagrams for cooking apparatus and domestic cookery, a section on the dairy and poultry yard, and a reprint of Nicholas Appert's 1810 treatise on the preservation of food. Cooper's advice to rural women with any physical complaints was exercise. Headaches, obstructed menses, immoderate menses—all could be cured in the same way—through exercise. He also advised countrywomen to keep accounts, to write down useful knowledge in books, and to take an interest in the general management of the farm. He told readers that consultations by husbands with wives would probably result in better management of domestic affairs.[23]

More important than Cooper in formulating the concepts of domestic economy was William Cobbett, the English political economist who lived in rural America for a time before writing his *Cottage Economy*, published in 1822 with a special American advertisement.

Cobbett was one of the most significant codifiers of the early nineteenth-century theories of agricultural development. After running away from his father's small farm in Farnham, England, at fourteen, Cobbett became successively a soldier, a journalist, and a major reformer; according to E. P. Thompson, he was the most important creator of a radical intellectual culture in early nineteenth-century England. While drawing his model of politics from agriculture, Cobbett concluded that the real condition of the working people was the test of political action and that it was laborers' right to have their needs met by the community. As Thompson concluded, Cobbett "nourished the culture of a class, whose wrongs he felt, but whose remedies he could not understand."[24]

Thompson did not analyze the importance of women to Cobbett's developing theories of political economy. Cobbett spent 1794 to 1799 and 1817 to 1819 in America, and during his first stay, he angered Jeffersonian Republicans with his defense of the British monarchy. After returning to England, however, he became deeply concerned with the place of women workers in the developing agrarian capitalism. Carefully observing agricultural workers, Cobbett became convinced that the condition of rural women workers was the final test of social conditions. After returning from his second stay in America, he wrote of workers near Gloucester: "The girls at work in the fields (always my standard) are not in rags, with bits of shoes tied on their feet and rags tied round their ankles, as they had in Wiltshire." This was, as Raymond Williams has noted, a radical shift of social viewpoint to a class viewpoint. No rural woman escaped to provide a distinctly female view of this life, but Cobbett provided for the laboring class what Jane Austen, in an area not far away, had developed for the middle class in her view of the importance of marriage in the shifting fortunes of rural families. Austen concerned herself with social improvement. For Cobbett, the test was more efficient work methods in all classes so that the lower middle class would not be eliminated and the workers sink to the level of the Wiltshire women, with their bits of shoes and rags.[25]

Cobbett intended his *Cottage Economy* as instruction for women of all ranks. He defined economy as management, which, when applied to the affairs of a house and family, had great importance because the character and ability of a people depended on it. The power of a nation, in turn, rested upon this ability and character. Cobbett saw women primarily as producers and as trainers of children to be producers. Servant women who could only clean, for example, were not as valuable as those who could bake, brew, milk, and make butter. Women who merely cleaned consumed food but did not produce it. Such producer skills were, according to Cobbett, more important than religion or even literacy. The American edition of this work particularly directed women to his precepts.[26]

It is within this tradition of the political economy of rural women that Hannah Barnard's *Dialogues on Domestic and Rural Economy* of 1820 must be interpreted. Barnard was the only American women to write about rural domestic economy during this critical transition in agricultural thought. Born on Nantucket off the coast of Cape Cod in 1754,

Barnard was part of that unique economy where the men left the control of domestic economy to women while they spent most of their time away fishing. When Michel-Guillaume-Jean de Crèvecoeur visited Nantucket in the 1780s, he was startled by the freedom with which women conducted the business of the island community. By the time of the Crèvecoeur visit, Hannah Jenkins had already become a Friend, married Peter Barnard of Hudson, New York, a widower with three children, and left the island. Barnard, according to Quaker tradition, was as a minister "an extraordinary woman for understanding and the gift of eloquence." In 1797, she and Elizabeth Coggeshall, a Rhode Island minister, visited England together, and there Barnard gave the first indication that she wished to develop her own interpretation of Quaker teachings.[27]

Controversy followed Barnard back to her Hudson Valley meeting. Quakers there accused her of holding erroneous opinions and called her before a special committee. Barnard provided a scholarly defense of her views, pointed out that similar opinions had appeared in books approved by the Society of Friends, and made the men uncomfortable by demanding that a friend Mary Macy be present rather than her husband. She called their opposition "very unreasonable" and insisted on circulating her opinions in book form. When she discovered that the spokesman for the meeting committee intended to render a verdict without reading her book, she told him he had insufficient ground to form a decisive judgment respecting its content. The meeting expelled her in 1802, after which, as one commentator noted, she "betook herself of useful domestic pursuits."[28]

Barnard remained in Hudson where she published her *Dialogues on Domestic and Rural Economy* in 1820. Cast in story form, Barnard's book was the first treatise by a woman on domestic economy and the only one of its time directed entirely to rural women. It showed how practicality and attention to productive skills would enable women to claim their place as equals in the rural household.[29]

In her *Dialogues*, Barnard told the story of farm women who, becoming solely domestic consumers, bring their family to ruin. Only the teachings of a skilled neighbor, "Lady Homespun," finally return the family to productive order. After the women of the Prinks family have been led into disastrous consumption by two wealthy women who settle nearby, Jenny Prinks, the daughter of the farm family, comes to Lady Homespun seeking a job doing fine sewing, her only skill. Jenny has no usable skills because she received no training in rural domestic economy. Instead of hiring Jenny to do fine sewing, Lady Homespun teaches her domestic economy—how to cook, wash, spin, weave, knit, and sew, as well as how to care for the health needs of the family.

Over the course of Jenny's apprenticeship, Lady Homespun trains her to be the perfect farm woman. The first principle taught is not to buy anything that takes trouble to maintain, carpets and extra furniture, for example. Then she teaches Jenny how to cook a calf's head and make headcheese. But, says Lady Homespun, "I shall give you no directions for that process [mock turtle], neither how to make puff paste, floating islands, figuring rolls of butter, cut into two, into the appearance of pineapples, not to eat, but fantastically to ornament a table; potatoe barras or nutmeg pudding, unmeaningly named after the smallest ingredient in the composition, or tricking off calf's foot jelly, or yellow flummery, with caroway comfits, etc. with fifty or a hundred more things, of the same tribe, which in my independent, and as I think, appropriate manner of affixing names, I call domestic follies." Lady Homespun advises using only cookbooks that are "well established on rational principles," and has Jenny write down the recipe for headcheese.[30]

Here is the moment of transfer of technology, of oral to written tradition, the secrets of housewifery made public. These secrets are being written down because the alternative written material is not appropriate for rural women. Lady Homespun then follows the same process in instructing Jenny on washing feathers from beds, bolsters, and pillows, and making candles and soap. Lady Homespun also refers with approval to neighbor women who make most of the family's clothes and produce enough surplus cloth to buy what they cannot produce, do their own work in winter, and hire help only in summer, at which time the help always eat at the table with the family and guests.[31]

Barnard develops an elaborate code of domestic relations for the farm couple. While the woman attends to her domestic economy, the man has a duty to do everything possible to make the work environment convenient for her. For example, Uncle Thrifty, who has arrived to help out the sinking Prinks family, announces that Jenny's mother, Harriet, must once again fulfill her duty as a "common farmer's wife ought to," making bread, sausages, soap, and candles, instead of reading novels. Uncle Thrifty brings several cows so that Harriet can tend her own dairy. He lends money to her husband so he can build a cowshed "to make it convenient for Ma to milk in bad weather," tells them how he has paved his cow yard at home with thick flat stones, and says, "Men ought to do their duty, in rendering every thing as convenient for their wives as possible." Barnard does not espouse complete equality. "Every family ought to have a head," she admits, "but that head ought frequently to commune with the heart which belongs to it, in order to ascertain whether what it directs and enjoins, will bear an impartial comparison with the golden Christian standards." Thus, a man should not be authoritarian but benevolent within his family.[32]

The daughter has another task to fulfill, that of obtaining a proper education. In addition to teaching domestic economy to Jenny, Lady Homespun also teaches her what she will need to be an educated rural woman. She must read geography, history, and "useful biography," learn to write and do arithmetic. When Jenny has mastered these subjects to her satisfaction, Lady Homespun rewards her with a book by Lucy Aikin, "Epistles on the Character and Condition of Women." Then she quotes a poem from Aiken:

> "Rise," shall he cry, "O Woman, rise! be free!
> My life's associate, now partake with me:
> Rouse thy keen energies, expand thy soul,
> And see and feel, and comprehend the whole,
> My deepest thoughts intelligent divide;
> When right confirm me, and when erring guide;
> Soothe all my cares, in all my virtues blend,
> And be my sister, be at length my friend."[33]

Here, then, is a complete political economy for rural women, combining domestic economic and social relations in a new way. Although there is no way to tell how many women read Barnard's tract, it expresses most fully a female ideal of rural economy in the early nineteenth century.

Like almanacs, the agricultural development literature emphasized women managing domestic production both for use on the farm and for the market. A concern that children be trained to this economical life led authors to define women's role more clearly. Mothering thus also took its place beside female industry and household manufacturing as crucially important. But, although the literature as a whole yielded to a greater emphasis on mothering and a narrowing definition of domestic economy in the 1840s, it seldom depicted a full-blown ideology of domesticity.[34]

The new agricultural development literature that flourished in the late 1830s and 1840s always provided contradictory advice to women. Editors encouraged in domestic occupations a wide variety of tasks that included producing both for family use and for the market. Lack of attention to either could mean misery and failure for the family. Farm family wealth still depended on domestic economy and that depended on women being industrious and economical. Rural advisers began to suggest in the 1830s that farmers subscribe to newspapers, and those closest to town and wealthiest no doubt read both newspapers and journals. Still, the messages remained mixed. Not until the late 1840s did the message of the domesticity that was developing in urban areas begin to be published in the rural journals. By far the strongest message was "The Whole Duty of Woman" that *The Plough, the Loom, and the Anvil* published in its first volume in Philadelphia in 1848. Women were advised to be complacent, elegant, frugal, and chaste. In the same volume, however, editor John Stuart Skinner advised male farmers to regard their wives not as higher servants but as intellectual partners and moral influencers. Wives were to influence morally, in his estimation, by reading natural history rather than women's literature.[35]

As rural men established public forums for educating themselves and exchanging information, women developed no parallel institutions to broaden their contacts. Agricultural societies and journals were run by men for men with only minimal attention to women's work, needs, or desires. Many of the early editors of farm journals were actually urban men like Skinner who knew little of farm women's work, nor did the urban women who published the new women's magazines, such as *Godey's Lady's Book*. Thus, farm women had no mechanism for exchanging their ideas or thoughts.[36]

Eventually, as Sally McMurry has pointed out, a form of "rural domesticity" did develop. But the term *domestic economy* was usually not narrowly limited to the house but rather included important farm production like raising poultry, dairying, and sometimes growing orchard and vegetable crops. Even farmhouse design reflected this productive orientation. Plans for the ideal farmhouse, published in agricultural development literature, featured the kitchen as the most important room in the house and tied it closely to the outside work areas where a woman's productive labor also took place. Parlors, those spaces that mediated between the family and the outside world, never came to be accepted in rural literature as important symbols of female space. Rather, they were family recreation spaces, or, as they were often called, sitting rooms. Even *Godey's Lady's Book*, that arbiter of urban female domesticity, recognized the need for informality in country living and the inability of country women to maintain the ordered precision of an urban parlor.[37]

Rural publications reflected the ideology of rural women's lives in ways that we still cannot describe precisely. Some Mid-Atlantic upper-class rural women, like the du Ponts, made the transition to an urban model of domesticity during the 1830s. Mary Johnson has described the contrast between the life of Sophie du Pont and her oldest daughter, Victorine, who lived on the banks of the Brandywine where E. I. du Pont established his gunpowder factory. Sophie had grown up in rural France where she learned the skills of operating a large farm household. While her husband was away on business in Paris, she supervised the farmhands during the hay and wheat harvests, oversaw the care of livestock, collected rents from tenant farmers, managed the fall vintage, and arranged for transport and sale of farm produce and butter to Paris. With the help of her brother Charles Dalmas, Sophie organized her Brandywine farm in much the same way after she settled in the United States. The farm provided some of the household necessities and some surplus for sale, while the powderworks accounted for the cash income that built the du Pont fortune. Sophie also managed a large ménage, offering services to some of the workmen as well as raising a family of eight children.[38]

Victorine, left a widow after a brief marriage, took over as mistress when her mother died in 1828 and remained in that status until 1837. The farm continued to provide some ham, milk, and butter after 1828, but rather quickly the du Pont household came to resemble that of an urban household. Victorine purchased flour, eggs, butter, lard, cheese, beef, and veal from local farms and small shops. Instead of purchasing material for clothes and making them at home, family members visited tailors, milliners, and mantua makers in Wilmington. The younger women did fancy sewing—ornamenting collars, embroidering, and quilting. Whereas Sophie had seldom left the farm, Victorine regularly shopped in the markets and stores in Wilmington and made trips to Philadelphia to purchase textiles, furniture, carpeting, housewares, and books. By the 1840s, virtually all the household necessities were purchased at shops. Dairying was done for experimental purposes only. Victorine turned to evangelical Christianity, supervising the Brandywine Manufacturers' Sunday School, and became part of what Johnson calls the "silent sisterhood" who subscribed to the ideology of domesticity.[39]

Nearby Quaker Hicksite women did not move toward the ideal of domesticity. Instead, they preserved Barnard's ideal of rural domestic economy, of women as producers and associates in the farm enterprise. When that seemed no longer an adequate ideology to describe rural women's conditions in the 1840s, Quaker women moved to a feminist critique of the doctrine of spheres, wherein women were to inhabit the private sphere while men occupied the public sphere. Then farm wives demanded the right to both economic and political equality. In many ways, farm woman Esther Lewis exemplified that transition.

CHAPTER 8

Esther Lewis: Biography of a Farm Woman

One day in 1827, while walking around her farm, Esther Lewis noticed that an area of the ground had loose surface stones of a peculiar color. She remembered seeing similar stones before—at an iron ore mine—so, picking up several samples, she sent them off to be assayed in Baltimore. She was right. It was hematite iron ore, and her Vincent Township farm was soon the site of an active iron ore mine.[1]

Esther Lewis was forty-five that spring, a widow of three years with four daughters, Mariann, Rebecca, Graceanna, and Elizabeth, aged eight, seven, six, and three. A son had died an infant, her husband four months later, leaving her the sole owner of their 150-acre farm. A relative of her husband, Joseph Lewis, yeoman, brought suit against Esther immediately after John's death. He contested the will because, according to family tradition, he did not believe she should have control of the farm and be trustee for her daughters. A nasty legal squabble with this relative dragged on for sixteen months with the case finally being taken to a jury of her peers.[2]

Esther emerged with her legal title intact, determined to manage the farm and raise her four daughters herself. She became the owner of a successful iron ore mine, used her farm as a refuge for runaway slaves, and educated her four daughters, launching one of them on a career that would bring her fame as an outstanding nineteenth-century naturalist. Moreover, she kept a detailed diary and maintained an extensive family correspondence that are an extraordinary record of a Chester County farm woman in the 1830s and 1840s. Esther Lewis assuredly was not typical. Yet her daily farm routine, recorded carefully in her diary, included the chores that the average Chester County farm woman performed. Although the ideal of true womanhood began its urban ascent in the decade of the 1830s, Esther's life and her records offer evidence that the practice of rural women was still far different from the urban vision of domesticity.

Except for Esther's discovery of iron ore on her Vincent farm, we know few details about her life before 1830, when she began her diary. The eldest of eight children born to Bartholomew and Rebecca Fussell in Hatboro, Montgomery County, she lived at home until marrying at thirty-six. In an effort to provide an equal inheritance for all his children, her father, a Quaker minister, moved to Vincent Township and then to three different farms before settling in Maryland. There Esther taught in a Quaker school near home while she waited eight years for her future husband, John Lewis (who also taught school and worked as a surveyor), to recover his health before marrying in 1818. A few months after their marriage, the couple moved back to live with John's Welsh parents on the family farm in Vincent Township. There Esther gave birth to five children in rapid succession. Five deaths came rapidly as well. Her father-in-law, mother-in-law, sister-in-law, and infant son all died in 1823, and her husband died in February 1824, a little more than two months after the birth of her fourth daughter.[3]

By the time John died, they had accumulated a small estate—over eight hundred dollars in bonds and notes and over one thousand dollars in household furnishings, farm equipment, and animals. The inventory taken immediately after John's death froze in historical perspective the interrupted farm cycle. There was grain in the ground, corn in the crib, rye unthreshed, flax unbroken, and the utensils of his work: a Dutch fan, cutting box, dung forks, rakes, and a hook, wheelbarrow, cow chains, two plows, corn harrow, common harrow, horse gears, a wagon screw, and a plantation wagon, hay (probably in the barn), along with rye straw, half a ton of plaster, half bushels and a basket, eighteen hundred feet of lumber, some calfskin, and tar pots. There was also a riding carriage, cart whip, his saddle and bridle, four horses, and one yearling colt. It was not a large farm, but they had seventeen sheep, three sows with their fourteen pigs, eight shoats, a large boar, seven steer, and six cows. The harvest that fall had been substantial, for they still had 114 bushels of corn, 84 bushels of oats, 46 bushels of rye, 7 bushels of buckwheat, 3 of wheat, 1 of flax, and a half bushel of beans in storage. In another storage room, perhaps the cellar, were other tools: a crosscut saw, stone sledge, grindstone, scythes and cradles, an ash hopper. There the tools of the yard and house intermingled: meat tubs and meat, lard tub, potatoes, bacon, tubs and barrels, half-processed flax, flax hatchels, flax brake, a loom and tackling, a churn and tubs.[4]

One can rearrange the goods in approximate order in the two-story stone house that still stands. The floor plan was a traditional four-room house, built in 1809, differing from Georgian houses of the time primarily in its two front doors. One entered through the left-hand door into the kitchen and through the right-hand door into the parlor. Each of these rooms had a fireplace, one a large open kitchen hearth, the other a double fire that served both parlor and bedroom behind. The left back bedroom had stairs leading up to a second floor, divided into three or four more rooms. John and Esther and their children had occupied two upstairs rooms, sister Mary one back and one upstairs room, John's parents the downstairs bedroom. A mentally handicapped brother probably occupied the garret. Into these small rooms, the Lewis family crammed all the tools necessary for use and processing of farm products.

The kitchen was simply furnished. There was a bake oven and dough trough, a brass kettle, a cupboard, table, stand, salt box, three pots, and andirons. The parlor was also simply furnished: a table, looking glass, bureau, child's cradle, cupboard furniture, ten chairs (six Windsor, four rush bottomed), andirons, shovels, and tongs. There was only one candlestick. However, the floor was carpeted, an amenity at the time, and there were window curtains.

The downstairs bedroom had similarly utilitarian furnishings. In addition to featherbeds and one chaff bed, the family had a bureau, a case of drawers, and six chairs. There were, however, ample linens, blankets, coverlids, sheets, tablecloths, pillowcases, towels, and linens. In the upstairs rooms were more featherbeds, sheets, towels, two pillowcases, yardage and sheeting, tow linen, flax, woolen and cotton yarn, and a reel. In one room was a chest and a clothespress. Tucked away in various rooms were feathers being collected for a featherbed, clover and timothy seed, flax and wool, bags, books, a map, desk and bookcase, trunks, tables, looking glasses, seven more chairs, a small stove, two umbrellas, saddle bags, and Esther's sidesaddle. It was a crowded household, filled with functional items and little evidence of luxury or affluence.

Esther had little more than the farm itself to support her family. Her responsibilities, including a dependent relative of her husband's to care for, would have quickly eaten up the small accumulated capital. Her mind must have been constantly at work on ways to make the farm yield a living for herself and her dependents. Finding the hematite ore was the first step toward the security the middle-aged woman sought. With hard work and help from kin, Quakers, and neighbors, she could now hope to survive. Yet the blow of John's death was a shock—even though he had been in poor health before their marriage. After his death, Esther never talked about him and the children learned to keep silent, too. He was remembered as a man whose chief aim in life was to do good. Esther carried on that tradition, but she had to combine it with an active life managing both field and household.[5]

Only the women's monthly meeting minutes of the Uwchlan Quakers record Esther's activities for the first six years after John's death. Within two weeks, Esther's name begins to appear in the records regularly—as representative to quarterly meetings, preparing certificates of removal, presenting testaments of disownment. In the Quaker separation of 1828, Esther stood firmly with the Hicksites, showing up as their clerk even before her own certificate of disownment was presented by her more orthodox neighbors. Esther did not go alone; other women of the Lewis and Fussell families in the area went, too. The next year Esther began her diary. She referred later to her writing as "little more than . . . the occurrences of the day, for the sake of reference in part to make amends for a treacherous memory (which I find very convenient in transacting the business allotted to me)."[6]

When Esther's diary opens, the farm was already an intricate web of family and work relationships, tied to neighborhood and meeting by other economic and social networks. To disentangle the skeins of social webbing, it is necessary to look at each: household relations, work performed, neighbor and community networks. The household of the Lewis farm in the 1830s was an extended one with kin, servant, hired laborer, tenant family, and orphans moving in and out. Esther's brother Solomon and sister-in-law Milcah, who arrived at Vincent in 1822, helped with the main farming tasks and moved into a new northeast addition in 1831. They had five children, four girls and one boy. When they left in 1834, a tenant farm family moved in for a year, but Esther found them "dishonest, vulgar and profane" and sent them away. She then asked Quakers Ann and Norris Maris to join the household as tenants. Ann and Norris remained for over twenty years. They came with one daughter Elizabeth after the death of three children in the previous seven years. While there, Ann gave birth to Mary Jane, Phebe, George Lewis, Debby, and Ellis. They belonged to Uwchlan Meeting, joined in antislavery work, both legal and illegal, and became an important part of the Lewis farm. Esther's parents also joined the household sometime in the early 1830s.[7]

In addition, Esther continued to care for Abel Lewis, her mentally handicapped brother-in-law, and brought a mentally handicapped sister-in-law, Hannah Lewis, home from the Quaker asylum. In one letter, Esther estimated that the two kin had to be waited upon thirty-nine times that day. So dependent was Abel that someone had to get him drinking water. Hannah escaped one evening and was found asleep in the orchard. She broke windows, ripped up sheets, pounded on the stove at night, and required constant attention. Although John Lewis, Sr., had provided a small legacy for their care, Esther boarded them more cheaply than could the Quaker asylum.[8]

Non-kin lived on the farm as well. In 1832, a family of three began boarding with them and Esther took in an orphan whose parents died in a flood in 1839. She often had live-in summer help and sometimes help who boarded in winter as well. She managed and organized the feeding of hired men, which, she noted in November 1832, consisted of six men and two boys with two more expected. During the harvesting of lime in April 1833, she counted ten men and boys at dinner. Another period when she was building a new springhouse and straw house, she had from three to thirteen workers there each day. Still, she wrote confidently to her sister in 1833, "though I have an unreasonable quantity of care on me, up stairs and down, In doors and out, I seem enabled to bear it. I have a good girl and that makes some of the rough places smooth, but you may depend she does not eat much idle bread with such a family. . . . I do what I can but as usual am called off every few minutes one way or the other."[9]

It is difficult to envision the crowded household in 1836. The Maris family occupied the kitchen and upstairs bedroom addition. Esther's parents occupied one back bedroom, she and the four girls another. Abel and Hannah must have occupied upstairs rooms, the hired girl perhaps the garret. In 1847, Esther made a list of the household. At that time there were fourteen members from four families: Lewises, Fussells, Trimble, and a William Edward. It was no wonder she wanted to build a larger house in the 1840s.[10]

The shift from a smaller to a larger house was made very consciously by Esther, as her household filled with various dependents and workers. She not only enlarged the size but rearranged it. She moved the kitchen to the rear of the house, facing the barnyard, enlarged it to provide more work room and planned to have running water. Off the kitchen was a dining room. She eliminated the parlor (which seemed a waste as they never used it) and substituted a sitting room. She saw the new home as more functional for aged and ill people who needed space and privacy on the ground floor. She moved the stairway out of the downstairs bedroom into the front sitting room where people could reach the upstairs without having to go through other rooms. Upstairs would be bedrooms for the more active members of the family. Like other rural women of her time who were experimenting with rearranging internal space to meet their needs for greater efficiency, Esther conceptualized her space needs. Then she hired the men needed to build the new house.[11]

The ore mine provided a year-round economic activity for the farm. Esther shared its profits first with her brother and then, apparently, with Maris. During the four years she kept records of the ore hauled off the farm, from 1830 to 1833, their hired hand moved over 1,300 tons of ore from the mine, averaging over 450 tons a year. In 1834 alone, over 800 tons were hauled. Esther and her brother Solomon split the $1.12 a ton, giving her an income of $470.61 for 1834 and several hundred dollars for each of the preceding years. It was not yet a full-time activity, for even in 1834 the hauling stopped at harvest time when the five horses and wagon were needed to haul farm produce. At other times, the wagon and teamster might be at work hauling lime or lumber. Still, the work was fairly regular with over 3 tons being hauled a day, 5 or 6 days a week. For the year beginning May 12, 1830, the teamsters hauled ore 135 days. The following year they hauled 185 days. Little is known about the day-to-day operations of the mine, for Esther seldom wrote about them, only mentioning in 1839 that they had a pump thirty-six feet long to pump out water and wash the fine ore. After 1834 the daily hauling amounts disappear from her diary, for she apparently started keeping separate records for the mine. There is also no record of where the ore was taken, but several iron-smelting works lined nearby French Creek, and the largest nail factory in the country, Phoenix Iron Works, was in operation there after 1824. Once established, the mine was a primary source of revenue for the farm but not of concern to Esther. The day-to-day management of the farm was the focus of her interest and efforts.[12]

A unique and important aspect of Esther's diary is that it not only reports work of female and male adults but also the work of her aging mother and young daughters. Thus, we can reconstruct a work cycle for women of different ages in the same time period. Although the full burden fell upon Esther, hired women, and adult kin at first, she also had increasing help from her daughters. She sent them to school but expected them to learn how to keep a farm household as she and her mother Rebecca had done. She considered their training in rural domestic economy so essential that at one point she thought that her daughter Rebecca perhaps should turn down an offer to teach. That Rebecca went off to teach anyway is perhaps a measure of the importance Esther placed on the intellectual and social growth of her daughters. Let us look now at the work lives of Esther's mother, Rebecca, of Esther herself, and of her daughters, three generations of farm women.

Rebecca was seventy-nine when she moved to the farm, a rural woman with long years of hard work behind her in Pennsylvania and Maryland, raising eight children and managing the farm while her minister husband traveled. There is no indication that Rebecca ever worked outside the farm teaching as did her daughter and granddaughters. Yet she was an amazingly robust woman, attending the Friends' meetings regularly into her nineties, visiting neighbors frequently, reading, and, apparently, thoroughly enjoying her old age.

Rebecca's major task on her daughter's farm was spinning. Although spinning yarn at home for cloth was no longer necessary by the 1830s, great quantities of fine flax and wool yarn were still needed for stockings that were knitted by hand in most households. This became Rebecca's primary work task, to spin and knit stockings. Esther, delighted in her mother's productivity, wrote frequently to kin about Rebecca's work.

"Mother's wheel hums yet, though it is silent just now, as she is doubling thread," she wrote at one point in 1841. The quantities of yarn were large; the last batch of flax had yielded enough yarn for seven pairs, Esther wrote one day. Graceanna wrote to her sister how beautiful and fine the thread was that her grandmother spun, and the girls took yarn with them to school to knit. Esther estimated that in 1844 alone, her mother had spun thirty-three thousand yards of stocking yarn.[13]

Esther obviously took pleasure in allowing her mother to specialize in a type of work that she enjoyed and could perform with such skill at an advanced age. But she herself had to bear the major burden of the household chores. Textile processing was not one of these chores any longer. With the advent of factory cloth, Esther could move textile processing from a central to a peripheral place in her schedule. Although the home had a loom in 1824, it was not being used by the 1830s, for Esther had wool taken to a factory six miles away near Lawrenceville to be woven. She did buy a carding machine so that the wool could be processed at home, but except for her mother's spinning, the rest of the textile process occurred outside the home and she purchased large quantities of flannel, calico, and linen. The men seem to have taken care of the processing of the flax that supplied Rebecca's wheel. There is no mention of processing any textile fiber except for some dying of yarn and cloth.[14]

On the other hand, sewing increased. No records of buying any ready-made clothing for the family exist, other than the purchase of shawls, occasionally stockings, and shoes. All the sewing for the house was done at home by the women, including making frocks, petticoats, and other clothes. The women also quilted and prepared household linen. They sewed what may have been large quantities of clothing for black refugees, who often used her home as a place to exchange their distinctively southern garb for northern-style clothing before they moved on. The women also prepared clothing for the antislavery fairs that became popular among abolitionists in the 1830s and 1840s. The few items of clothing the Lewis family purchased came from the free produce fairs to which they also contributed.[15]

Food processing also increased in the 1830s. The Lewis farm bought coffee, tea, sugar, fish, spices, chocolate, molasses, and soda in the early 1830s as well as tobacco and snuff. Their flour was ground at the mills, but they baked all bread and pies at the farm. At Christmas in 1831, Esther counted forty-three mince pies baked in the outdoor oven. All the butchering of beef and hogs and processing of meat was done on the farm. They raised chickens, turkeys, and geese and saved feathers for featherbeds and pillows. As Esther produced more, she apparently expected the same of her poultry: in her recipe book was another farm woman's suggestion to use red pepper to make the hens lay all year. She processed large quantities of butter, preserved fruit, and made sauerkraut.

Esther also worked in the garden, experimenting with tomatoes and even hops. She supervised the drying of apples, the making of applesauce, and the rendering of lard. The women not only provided enough for their extensive household but also prepared large quantities of butter, eggs, poultry, lard, and meat to send to market along with oats, cider, apples, and lambs. They produced the large quantities of candles needed in the household, making, by Esther's count, one thousand in 1830. And they experimented with the new labor-saving "soft soap" for their weekly washing.[16]

But some of the farm income went toward the increasing purchases of consumer items for processing. They bought a new cooking stove for twenty-eight dollars in 1834 and a washing machine for nine dollars in 1840. They purchased earthenware and tinware, and items like broom handles, scales, and varnish. Although Esther often traded butter to local artisans for items like shoes, she used the cash from market sales to purchase other things in the city. Market day became a complex exchange of farm-produced goods for goods to make the farm an increasingly productive economic unit in a market economy.[17]

Because inventories of the Lewis farm exist for both 1824, the year John died, and 1848, when Esther died, it is possible to judge how much the farm moved from production to consumption. The difference in material culture between the two dates was very small. There were changes in field farming technology. In 1848, there were cultivators, a horse rake, and a corn sheller among new farm implements. Yet, even if Esther and Maris had access to more extensive field crop technology through cooperation with neighboring farmers, their investment in tools had been very modest. Nor did the number of animals change dramatically. Esther's half of the farm property came to less than $950.

In the farm household, only a few changes were evident. Except for the washing machine and cooking stove already mentioned, the additions were few. There was more tinware—fourteen tin milk pans indicated increased dairying—and there were buckets and other items. She now had seven candlesticks instead of one, indicating the household read and sewed more often. There were three pairs of clothes irons and more tubs, probably because more time was spent on maintenance of clothing. There were a few new kitchen utensils—more brushes, some dripping pans, a gridiron, an apple parer, some teapots, and a coffeepot. There were more cups, plates, and bowls. Only the washing machine, the cooking stove, and possibly the apple parer could be termed labor saving.

There were a few new amenities. Washbowls and pitchers had been added. Straw carpets, blinds, a clock, three rocking chairs, a settee, and cushions were new in the sitting room. There were quilts in 1848, whereas none had appeared in the 1824 list. Esther even had a few silver teaspoons and tablespoons.

Still, the change in material culture was not at all great in the twenty-four years that Esther managed the farm. Most items were of the kind available earlier to farm families. Probably the straw carpet, the rocking chairs, and the blinds were the only consumer items that were really new to farm households of the time. The new house had given a bit more privacy and space. Running water—if the new house actually had it—would have been a major labor-saving device. But household production, although more labor intensive, took few new tools. This lack of change seems to demand at least some reordering of the application of the ideology of domesticity to rural women. There may have been more productive work rather than less in these early nineteenth-century farmhouses, but not much more consumption.[18]

Esther also spent money on books. She did not consider these a luxury, but rather an essential expense. Her purchases are important because they offer a rare glimpse of what rural women were reading. There is no mention of any literature on domesticity in Esther's diary and letters, although she did have a collection of recipes, including one cure for English cholera that began with "a teaspoonful of laudanum." She also joined the Kimberton library and mentioned borrowing a book of receipts. But no other books that related to domestic affairs are mentioned in her diary. Instead of the literature of true womanhood, she nurtured herself and her children on antislavery tracts. She purchased a family encyclopedia, a geography, and an atlas, but beyond these basic reference books her main literature was abolitionist: *The Genius of Universal Emancipation* by Benjamin Lundy, the *National Anti-Slavery Standard* edited by Lydia Maria Child, the *Freeman*, William L. Garrison's *Liberator* and *The Non-Resistant*, Elizabeth Chandler's *Works* (which included exhortations to women to take responsibility for the abolition of slavery), and Theodore Weld's *Slavery As It Is*. In September 1837, Esther noted under purchases, "Grimké's Appeal," referring to Angelina Grimké's "Appeal to the Women of the South." Thus the market economy that linked the Lewis farm to the marketplace also linked it to the marketplace of ideas in which antislavery thought played the major role.[19]

Besides the literature of reform, Esther also purchased two other types. The first was Hicksite publications such as the *Friend's Miscellany*, the *Quaker*, and classics like George Fox's *Works*. The second type was the literature of the early nineteenth-century romantics. Except for Whittier's *Poetry*, little of this literature is now remembered, but it formed the underpinning for the union of classical Quaker thought and the newer antislavery thought. It included such popular books as Job Scott's *Work, Tale of a Grandfather, Ann Nature, The Orphans, History of the Jews*, and *The Life of Mohamed*. Whittier particularly delighted Esther. In a burst of emotion, she wrote of Whittier's "Judea": "How sublime. How superlatively beautiful." The books that came home in the farm wagon, then, seemed to combine the ideology of the late Enlightenment, early romantic literature, and reform thought.[20]

Esther's passion for reform extended to her expectations for her daughters as they reached their late teens. Apparently, the girls never learned to spin, but they did great amounts of sewing and were instructed in food processing. They went to a nearby day school before boarding at Kimberton school only a few miles away. Teaching of domestic economy went on amid a public education that Esther supervised carefully. All the girls had home duties and were instructed in "house business," yet Esther sent them to lectures and lyceums, and encouraged their political activity and their work outside the home as teachers. All four daughters taught school for some time in the 1830s. In addition, the girls while young helped with picking, whitewashing, delivering produce to neighbors, fetching mail from the post office, writing letters to kin, dairying, and generally assisting their mother. Esther seems to have been particularly conscious of a dual role for her daughters—both public and private. She had taught but never lived away from home herself and seldom attended public functions. Her daughters all attended boarding school, boarded away from home while teaching, and enthusiastically attended public meetings with their mother's encouragement. In their absence, Esther depended upon her hired girl but also accepted a great deal of the burden herself. The involvement of the mother and daughters in reform meant increased burdens for each member of the Lewis household.[21]

Part of that reform activity took place in the home with the hiding, clothing, and nursing of runaway slaves, an important part of Esther's self-imposed reform obligations. Her farm was an important link between Thomas Garrett in Wilmington and the city of Philadelphia. Most of the refugees moved across the Mason-Dixon Line from slave territory onto free but unsafe Pennsylvania soil and then through the Brandywine Valley. They traveled in farm wagons by backcountry roads to one of the safe houses, where they received new clothing and money for their passage to Philadelphia and the North. It is difficult to tell from Esther's diary exactly when and how many refugees she sheltered, but there are references to "strangers," and to taking "blacks" to the railroad at French Creek. There were enough refugees to make their presence a regular part of the Lewis farm routine.[22]

The general reform movement of the backcountry in which Esther and other rural Chester County women participated was a central and an absorbing part of the lives of all the women of the Lewis household. Her daughters taught at the schools established for refugees and noted their students' delight in learning. In the 1840s, Esther, Rebecca, and Graceanna were active members in the local Lundy Anti-Slavery Union, serving as officers and representatives to the county antislavery meetings. This reform activity gave meaning beyond the daily routine that Esther struggled through. Some portion of the profits of the farm went to support the reforms, a way of tithing to the community at large.

Reform activity went on amid a sometimes hostile larger community in the 1830s and within a relatively inactive group of Hicksites as well. Within the Hicksite country meetings, however, was a small band of antislavery advocates who saw their role as one of raising the consciousness of neighbors, friends, coreligionists, Pennsylvanians, and Americans to the need for change in the structure of American society to include civil rights for black citizens. This was not their only reform goal, but it absorbed much of the Lewises' energies during the 1830s and 1840s. In a sense, the most progressive Hicksites, including Lucretia Mott who frequently visited the Uwchlan Quaker meeting, were Esther's real political family. With them she was at home, joined in sentiments that did not seem radical but rather moderate. Esther was not able to participate publicly in the women's movement as it developed within the Hicksite progressives, but she formed part of the background for that movement. She passed her self-confidence and sense of social responsibility on to her daughters. They moved from antislavery to temperance to women's rights. When Esther died in 1848, the year of the Seneca Falls women's rights conference, she was part of an older generation that had laid the foundation upon which younger women could build. As Esther lived and breathed politics, she bequeathed it to her daughters along with their equal portions of the farm.[23]

Esther's brand of reform did not include any interest in changing the basic familial or economic structure of society. After her daughter Rebecca and son-in-law Edwin had lived in Indiana from 1834 to 1841, they wrote home that they were considering joining a farm commune. The response was speedy and unequivocal. Esther wrote that she and the other daughters all opposed it because the two would lose their independence and not be able to read what they wished.[24]

On the other hand, Esther did believe in the type of cooperation that could take place among neighbors settled on individually organized family farms. She formed a partnership with Norris Maris, and they purchased farm tools and animals together. She noted in her diary the help of neighbors at harvest time and of quilting with her neighbors. Both she and her parents visited non-Quaker Germans in the neighborhood. They also seem to have had good relations with neighboring Baptists. Her view of farm life was of cooperative but not collective farming.[25]

Certainly, Esther realized the declining ability of farm families to maintain their old mode of production as the country industrialized and urbanized and as farmland became more expensive. Yet her response was to create new year-round sources of income from the farm, such as the mine, and to expand market production, especially butter and poultry.

Esther also realized that western expansion was an essential outlet for part of the farm population. When her brother and sister-in-law Solomon and Milcah Fussell left for Indiana at the end of September 1832, she carefully recorded in her diary their names and ages along with those of their five children. The following May, she noted the departure for Indiana of her sister and brother-in-law, Ann and Jonathan Thomas, and their daughter, together with eight families. These families were composed of parents in their forties with half-grown children. They went west because they could purchase land more cheaply and reproduce in the West a farm family similar to the one that Esther and John had produced in Chester County.[26]

Esther also developed a sensibility to the beauty of nature that strikes one as another important dimension of her life. Farm women have seldom written of the natural beauty around them, and so some writers have concluded that rural women might have seen farm life only as a burden of incredibly hard work. Esther's writings show a response to nature that, at times, breaks through her attention to the details of farm management and reform activities. One day in February she noted turtle doves cooing, robins singing; on a day in April, the "sun unusually red"; and one July, "fine weather for harvesting." And she found beauty in winter as well. On a cold February in 1830 she commented on the sleet: "Every blade of grass and every pointed thorn seemed etched in glass." In January of 1832 she noted, "Rain froze as it fell and formed a most elegant sleet." The next month she wrote, "Trees elegantly dressed in sleet." On August 8, 1844, "a beautiful butterfly was found in the Store room this day, with dark spotted wings edged with pale yellow." Among the things she bequeathed to her daughters, then, was an attention to the natural world that surrounded her in the country.[27]

In the last two years of her life, Esther was confined to a wheelchair and to the front porch from which she wrote letters, jotted in her diary, and planned her new house. She noted vaccinations for smallpox in January of 1846, another visit of Lucretia Mott to Phoenixville in April. In May 1847, she wrote that the weather was cold and grain prices high, and that it would be hard on the poor. How to achieve international peace was a question being discussed in Phoenixville, she noted in her diary in October of that year. On January 29, she recorded receipts for subscriptions to the *Freeman* and the *Liberator* for the family and noted briefly on February 2, "My foot much deseased." She died on February 8, 1848.[28]

No other rural woman emerges quite so clearly from the documents of early nineteenth-century Chester or New Castle counties as does Esther Lewis. Farm daughter, early teacher, mature wife, widow after a brief marriage of six years, an ambitious discoverer of iron ore and mine owner, she organized an extended farm household, sheltered black refugees, read Whittier's poems with delight, noted the beauty of a butterfly's wings. Not an ordinary woman, but one who in her life summed up the richness and diversity of what life could be for a countrywoman practicing her rural domestic economy in the decades from 1820 to 1850.

Biographies of rural women can help explain change in ways that other material cannot. Although it was not ordinary, the life of Esther Lewis gives us a prototype of what the everyday life of women might have been. Lewis was able to adjust herself to the available options, to develop skills and resources that made it possible to function successfully even without a principal male in the family. Drawing on kin and religious community, she reknit the family after the deaths of several central members and successfully reproduced the farm family through her labor and the organization of a complex web of social and economic ties. She passed on what she believed to be fundamental attitudes to her daughters—self-sufficiency, work for the market, cooperative support networks, intellectual curiosity, concern for how the larger community cared for its members, support for political action to change policies. The life course of the Lewis family thus gives us one model of rural women beside which historians may begin to place others. If the family is the primary social institution within which human energy is produced and socialized, then biographies such as this one can give us insight into how that basic energy takes form in society.

PART III

THE PUBLIC SPHERE

CHAPTER 9

"Centre Then, O My Soul!" Ministering Mothers

Between 1750 and 1850, women of the Brandywine Valley became visible in the public sphere in three major ways—through ministering, teaching, and reforming. In each area, Quaker women were the most visible. Quaker mothers, almost alone among women, occupied official positions of religious leadership as ministers in the eighteenth century. Quaker daughters led the movement of young women into the public sphere as teachers early in the nineteenth century. By the mid-nineteenth century, Quaker women were appealing to the bonds of sisterhood in an effort to expand occupational and political rights. Although they were not the only women filling the roles of minister, teacher, or reformer, they took the lead in exercising and defending the right of women to enter these public activities. Because of the importance of the actions and arguments of Quaker women in the public sphere, it is necessary to understand in some detail the role of Quaker women ministers.[1]

To become a Public Friend, as Quaker ministers were termed, was an important step in loosening the bonds of womanhood. Quaker ministers were not formally ordained but practiced what is often called lay ministry. Antoinette Brown was the first woman ordained by a church that required formal ordination. That was in 1853 and few women were ordained in the century following. Hundreds of Quaker women, however, had served as ministers by the time Brown was ordained.[2]

Recent research also shows early feminization of the churches in nineteenth-century America, women's essential role in revivals through prayer meetings, and the existence of a considerable number of women who preached despite the official opposition of all sects but Quakers against it. The tradition of lay ministry increased rapidly between 1750 and 1850 and changed in important ways. Because major Protestant denominations refused ordination to women, they usually exercised their ministry in alternative ways. Many women became itinerant revivalists, others formed utopian communities, and some founded new churches. The black preacher Sojourner Truth is the best known example of the first; Ann Lee, founder of the Shakers, of the second; Mary Baker Eddy, founder of Christian Science, of the third. A large number of lesser known women followed this pattern as well.[3]

In addition, some churches recognized the practice of women preaching while refusing to recognize the principle. This recognition of a lay ministry within major churches allowed them to utilize the power of women's ministerial leadership without conferring authority. The practice began early and often involved explicit denial by these women of their right to preach. A number of women were allowed to preach regularly in pulpits of major denominations as long as they did not formally claim the right to do so.[4]

Another alternative for women was to move their preaching from the Sunday morning pulpit to another time or space. Leontine T. C. Kelly has pointed out the importance of the mid-week prayer service for black women where, "the pew became pulpit" and the unordained black women became "spiritual leaders." Female ushers, elders, and deaconesses in black churches often exercised important preaching roles. In early nineteenth-century Delaware, black churches licensed women as preachers but did not allow them to be ordained. Major Protestant churches in the 1880s similarly recognized the right of white women to become part of the deaconate as well as missionaries. Black women organized female prayer bands in the early nineteenth-century African Methodist Episcopal church. White women likewise organized women's prayer meetings in revivals in the early nineteenth century. They also held Bible classes, established Sunday schools, and took increased responsibility for evangelical conversions. Recent research documents the existence of an important lay ministry within Protestant churches for black and white women extending through the nineteenth century. Many additional women developed a literary ministry in the early nineteenth century by using written fiction as a vehicle for preaching.[5]

What made Quaker women ministers unique was their insistence on the principle as well as the practice of women preaching and their acceptance by the organizational structure of the Society of Friends. Although all Quaker ministers were lay ministers in the sense that they received no formal ordination or pay from the Society, by the end of the seventeenth century they received official recognition through certificates of acceptability issued by their meetings. Quakers claimed this simply recognized the gift of ministry, but through it women members moved into a public leadership role. They took their place with male ministers in raised seats in the ministers' gallery in front of the meetinghouse.

When a woman became a Public Friend, she then claimed a new place in the public sphere, with the official support and sanction of her religious community. Other Protestant male ministers, unlike the Quakers, interpreted the Bible as prohibiting women from preaching. A dialogue between forty-seven-year-old Ann Moore, on a ministerial trip through the Mid-Atlantic in 1757, and a Presbyterian minister reflected the Quaker insistence on the right to preach. Moore, already nineteen years a preacher, reported this exchange:

> After this Meeting as we sat at Dinner, he began to vent himself, asking me where I lived; I civilly told him; have you, said he, read Paul's Works? I answered, *Yes*; well said he, and what do you with that Text Ma'm, or dont you choose to meddle with it? I told him I took it as it was; if he had said so in one Place, he in another recommended several Women as Fellow Labourers in the Church: and the Prophet spoke of the Day that was to come, when the Lord would pour out of his Spirit upon Sons, and upon Daughters, on Servants and Handmaids, and I did not think it was reasonable to suppose that the Scripture contradicted itself.[6]

Because of their insistence on this principle, women Public Friends occupied a position different from other lay women preachers. But that prominent position within the Quakers and in early American society also changed over time. Whether as seventeenth-century martyrs, eighteenth-century pioneers and settlers, or early nineteenth-century reformers, Quaker women ministers reflected an increased female presence within the society and in the world outside.

The stories of the lives of martyrs became an important source of role models for later Quaker women and girls. These pre–1700 martyrs were a small group of women, most of them born in England, who helped create a Quaker presence in the New World and fought for the right of Quaker women to preach in public in the New England colonies. The final acceptance of female ministers in New England came only in 1704, after over a half century of debate that began with Anne Hutchinson's trial in 1636. In addition to winning external acceptance, these early women firmly established and defended the principle as well as the practice of women preaching within their church, of women having separate business meetings, of women's self-selection to the ministry and for ministerial visits, and of married women ministers traveling without their husbands in these visits. The range of the reforms is impressive.

The external battle for the right of Quakers to preach in which Quaker women participated so actively gave them a special place in Quaker history as martyrs. English historian Margaret Spufford has pointed out that liturgical change in England produced a fringe of seekers in seventeenth-century rural villages like Cambridgeshire. Such spiritual seeking was extremely widespread among women and girls. Converts to dissenting sects were predominantly women; the majority of Quakers arrested for religious dissent were also women. Although Quaker males remained in control of this growing body of dissenting women, from the beginning women were attracted to a faith that needed no scriptural authority for practice.[7]

Women became active missionaries in Germany, Britain, and America. Frederick Tolles has estimated that almost one half of the Quaker missionaries who arrived in America between 1656 and 1663 were women, as were four of the eight Quakers who arrived on the *Speedwell* to urge Bostonians to repent in 1656. A 1650 to 1700 meeting record from early Newport, Rhode Island, listed one-third of visiting Quakers as women. Few are designated in Quaker records as ministers, however. Two major sources exist, one a catalog complied by J. W. Lippincott of all deceased ministers and elders, and the second a collection of biographies by Willard Heiss. The Lippincott catalog begins with the entry "Mary Dyer, hanged, 1.14.1660," but it lists as the first woman minister Elizabeth Hooton, the English Quaker whipped eight times in Boston, and gives only one other woman as minister during this period.[8]

Although not yet officially designated as ministers, these early women—Ann Austin, Mary Fisher, Mary Dyer, Ann Burden, Mary Clark, Elizabeth Hooten, and other women from Rhode Island—suffered jailing, whipping, and in Mary Dyer's case, death for demanding religious toleration for Quaker tenets. Two other women, Deborah Wilson and Lydia Wardell, protested the treatment of these Quaker women by stripping to the waist and appearing in the churches of Newbury and Salem. Margaret Brewster went into Boston's Old South Church in sackcloth and ashes. One Quaker historian later labeled these actions "hysterical tendencies." Yet the religious commitment of women and these dramatic actions were the rhetoric and metaphor of a larger protest. By 1680, all Quakers had successfully gained the right to hold meetings in Boston, but they did not win the final acceptance for women's right to preach in Massachusetts for another twenty-four years. In 1700, a Boston mob interrupted a woman in Elizabeth Webb's party who was attempting to speak outdoors. An English Quaker male who was in the party had to plead for freedom of speech before the crowd was quieted. Esther Palmer and Mary Lawson, who visited Nantucket a few years later, apparently witnessed the last opposition to Quaker women preachers. As one chronicler said of their visit: "They were living ministers and their testimonies reaching and affecting the people, all objections were removed against women's preaching, without the labour of dispute, contention, or jar about it."[9]

The vehemence of the opposition directed specifically against women preaching resulted in its first formal published defense by a Quaker woman in 1666. Written while she was imprisoned in Lancaster Castle for refusal to take the oath of allegiance to the king, Margaret Fell's basic argument was for the spiritual equality of men and women, an equality manifested by women preaching.

In an exegesis that Ann Moore and later Quaker women ministers were to rely upon for their defense, Fell carefully reviewed all biblical references to women speaking and prophesying. She concluded that there was ample evidence to support women speaking but that for over twelve hundred years "Apostacy" had kept the true light from shining. "And so," she argued, "let this serve to stop that opposing Spirit that would limit the Power and Spirit of the Lord Jesus, whose Spirit is poured upon all flesh, both sons and Daughters." In a special rebuke of the paid ministry, Fell criticized them for using the words of women in the Bible and then denying women the right to preach: "You will make a Trade of womens words to get money by, and take Texts, and Preach Sermons upon Womens words; and still cry out, Women must not speak, Women must be silent; so you are far from the minds of the Elders of Israel, who praised God for a Womans Speaking." Fell insisted that in the "True Church, sons and Daughters do Prophesie, Women labour in the gospel," and these true speakers, men and women, would triumph over the "false Speaker." Women ministers depended upon Fell's words not only in personal bouts with complaining ministers but also in preaching about the positive role of women in opposing evil.[10]

It may also have been the continued harassment of women preachers and the need to discuss ways to respond to it that led women in Boston to hold separate meetings in 1672. London Quaker women had established separate meetings as early as 1656, but these were primarily to handle social welfare. When Fox arrived in New England in 1672, he sanctioned the meetings already begun and encouraged other women to do the same, writing a defense of women's meetings four years later in which he argued that women needed to speak among themselves of "women's matters." Part of his concern, and perhaps of the Boston women as well, was to discourage "disorderly walkers" such as Jane Stokes who had earlier opposed all form and ceremony as well as regular meetings. People called ranters, who believed God existed in all creatures and who sometimes practiced ritual nudity, seemed to be particularly endemic in the northern colonies and were probably an extreme form of the antinomian beliefs that Anne Hutchinson espoused earlier. The encouragement of women's meetings where discipline could be handled by women probably resulted in the loss of some women to the Quaker faith, but it gave those remaining a more secure role in their own governance within the church. With that change, the development of a regular female ministry was assured, a position of authority that other women of the reformation lost as their congregations became more stratified.[11]

During the 1680s, other meetings followed Boston's lead. Philadelphia Yearly Meeting established its women's meeting in 1681. By the end of the century, women had a place in the organizational structure of the Society of Friends. As Mary Maples Dunn has noted, these meetings did not give women equality with male meetings. They did, however, give women official time and space to discuss their common religious concerns. It was a major innovation for a seventeenth-century church. In the next century, American women would work with English women to establish a similar yearly meeting in London.[12]

Quaker women thus gained a commitment from the Society to support their preaching and assemblies. By the end of the seventeenth century, women had also established their right with the men to self-select themselves for the ministry, to make ministerial visits, and to determine when they should travel. While the community undoubtedly exercised considerable influence on whom they recognized, there was no further official qualification other than that the woman felt called to the ministry and that her local meeting considered her worthy to represent it. Once accepted, the woman might still find her meeting arguing against her travel, but if she persisted in her "concern" to visit and if the meeting approved that visit, the approval could take precedence even over her husband's wishes. In a least one case in the late seventeenth century, a woman sought a certificate to visit other meetings over the opposition of her husband and received it. Jane Biles presented her case in 1699, and as Heiss reported it, the meeting "not being satisfied with her husband's opposition" left her at liberty to go. He then asked to go along, and after some time, the meeting agreed to allow him to accompany her. In another 1699 case, Ann Dilworth's husband also asked for a certificate to accompany her, but the meeting in this case turned him down. Thus a married woman could negotiate a considerable amount of physical mobility as well as space within the Quaker structure.[13]

By the end of the seventeenth century, the right of Quaker married women to travel long distances in the company of women companions and of men not their husbands had become firmly entrenched. It was a right they would exercise increasingly during the eighteenth century. Quaker women ministers became a common sight as they crossed the Atlantic and criss-crossed the colonies in an increasing number of visits. The memoirs of eighteenth-century ministers and of their women's meetings are studded with accounts of a woman having "drawings on her mind" or "a concern" to visit a neighboring meeting, one in the next colony, or England. During the eighteenth century, almost 40 percent of all American Quaker ministers visiting England were women. Visits to

American meetings were also frequent, although perhaps not as much as they had been earlier. Twenty-seven percent of visiting ministers and elders at Newport, Rhode Island, were women in the eighteenth century, a percentage that climbed to one-third in the early nineteenth century. Purchase, New York, had almost 50 percent female visitation.[14]

According to women who kept careful accounts of their travels, usually by horseback through the backcountry to isolated meetings, the distances covered were enormous. Although some women never left their own meeting or visited only other nearby meetings, others seem to have had a frequent "concern" to visit, often returning from a long trip only to embark shortly after on another. Jane Biles traveled through the Mid-Atlantic in 1689, to New England in 1696, to Great Britain in 1702, and back through the Mid-Atlantic in 1704. Elizabeth Teague traveled to the South in 1720, through the Mid-Atlantic in 1721 and 1724, and back to the South in 1725. Ann Parsons went to New England in 1705, to Great Britain in 1709, back to New England in 1713, and to the South in 1714 and 1719. Elizabeth Whartnaby traveled to the Mid-Atlantic in 1715, to the South in 1717, to New England in 1718, to the Mid-Atlantic in 1726 and 1729, and to the South in 1728 and 1729. Susanna Morris went to the South in 1722, through the Mid-Atlantic in 1728 and 1732, to the South again in 1734, to New England in 1742, and to Great Britain in 1744 and 1750. Ann Moore logged over a thousand miles and ninety-one meetings in 1756, and Elizabeth Welkman, an English minister, over four thousand miles in 1761–63. As Ann Chapman Parson said in her last illness: "I have traveled a pretty deal in my time."[15]

Whether as active traveling ministers or mainly as a presence in their home meetings, the number of women ministers and their proportion of all Quaker ministers grew during the eighteenth century. Although the Philadelphia Quarterly Meeting admitted women ministers to its general meeting of ministers in 1718 and other branches of the Philadelphia Yearly Meeting soon followed the example, in the first half of the century only approximately 10 percent of the ministers were women. By the end of the century, however, as many as 50 percent may have been women. Forty-two pecent of ministers dying in the second half of the eighteenth century were women; the number reached 50 percent by 1800–29. Not only the Philadelphia Yearly Meeting showed this increase; the London Yearly Meeting did, too. By 1828, there were almost ninety women ministers in the Philadelphia Yearly Meeting alone.

That year forms a great divide in Quaker history and also in the history of women ministers. The Society of Friends fought its way to an acrimonious split that left two factions—Hicksite and Orthodox—both claiming to represent traditional Quakerism. The split seems to have originated in two opposing responses to religious and social changes of the opening decades of the nineteenth century. Both sides were deeply concerned about the deism and worldliness of society. The Orthodox solution propounded by the urban church elders in Philadelphia had adapted evangelism and a new emphasis on scriptures as a method of restoring faith. Elias Hicks and his followers, on the other hand, supported a reemphasis on the "inner light" as the sole basis of spiritual authority that had led Quakers to their original position as English dissenters. Hicks disapproved of Bible societies, study of the scriptures, and the other institutional trappings that had accompanied the Protestant revival of the 1820s. His followers argued that his interpretation allowed for greater freedom for meetings and for individuals within them.[16]

There is no way to tell which way women leaned in that painful separation, as no statistics were kept on the number of ministers who left or their gender. Lucretia Mott was only the best known of the female Hicksite ministers. Alice Wilson, an influential rural minister at Centre, Delaware, also left. Meeting minutes reveal that women ministers played a role of some importance in the separation. Most of the Brandywine Valley Friends became Hicksites. Historians have still not precisely identified the social basis for the doctrinal disputes that divided Friends so deeply. What is certain is that the hinterland was disaffected socially as well as religiously. Urban Orthodox Friends seemed to rural Hicksites to be abandoning the traditional Quaker way for a more evangelical religion. At the same time, once separated from their Orthodox coreligionists, Hicksites moved quickly into progressive secular reform circles, the ones that the Grimkés felt comfortable within. Mott found her spiritual home with the rural Hicksites rather than the urban Orthodox Friends.[17]

As more Quaker women became ministers in the late eighteenth century, they also increased their presence by simply living longer. Seventy women ministers about whom some biographical information is known died between 1700 and 1749 and 143 died between 1750 and 1799. The typical early eighteenth-century minister married at thirty-two, entered the ministry at thirty-five, remained a minister for twenty-five years, and died at sixty-two. At least four-fifths were foreign born. The typical minister of the late eighteenth century, on the other hand, married at twenty-six, entered the ministry at thirty-three, remained a minister over thirty years, and died at sixty-seven. At least two-thirds were native born. European marriage ages were usually higher than those in the colonies; thus it is not surprising that the native-born women would marry at a younger age. Later, Quaker marriage ages again increased (see Appendix table 18).[18]

Many of these eighteenth-century women ministers wrote journals, thus participating in the Quaker literary tradition. These journals, or autobiographies, are significant both for the differences between Quaker and Puritan women, and the differences with Quaker males. Carol Edkins argues that although journals for both Puritan and Quaker women "symbolically celebrated their sense of community via the written word," Quaker women were in rebellion against Puritan norms, whereas Puritan women supported an exemplary rather than a rebellious life. Quaker women also shared roles as ministers with Quaker males, whereas Puritan women had separate roles as wife or mother. Both Quaker and Puritan women in writing journals and offering them to the community were moving toward the public sphere. But Quaker women more often broke with their families in joining the Quaker community.[19]

Female and male Quaker writers did share common characteristics. As Howard H. Brinton has pointed out, most avoided discussing personal experiences except those that concerned religion. Few mentioned the subject of sermons or theology. Many recorded stages of spirituality in which divine revelations that occurred in childhood gave way to youthful playfulness later recalled as a waste, only to be followed by a divided self and preparation to devote one's self to following the light. Finally came the first speaking in meeting and the adoption of plain dress.[20]

These stages can be discerned in many of the women's journals dating from the period between 1750 and 1850. There were, however, additional social aspects of the ascent to spirituality that seem particularly important for discussion of these women's lives. What types of families did the women emerge from? Did the women have any difficulty in accepting the call because they were women? How did they handle their traditional family duties after becoming acknowledged Public Friends? Did they receive support and encouragement from other male and female ministers in developing an appropriate ministerial style? Did women seem to use a particular imagery in their speaking and writing? Did they have a consciousness of their peculiar role as females in a traditional male role? Finally, what was their influence on and relationship with other women and men within the church and outside? Each of these questions can be answered only in part. There are, however, clues in the journals, in accounts left by others, and in a few personal letters that enable one to understand the ways in which these women dealt with their public roles in an era when they were almost entirely alone as women in their occupancy of public space and authority.

A number of the most fascinating childhood accounts are by women who came, as Susanna Lightfoot said, from "limited circumstances." In most of these cases, fathers died at an early age, and the young girls were put out to service; their stories chronicle their difficulties in reaching the ministry through economic as well as spiritual struggles. Lightfoot, put out to service with a Quaker woman minister, served as her personal maid but also tended cattle and horses during ministerial tours. Jane Hoskins, whose father disowned her for marrying someone not his choice, indentured herself to pay her fare to Pennsylvania in 1712. She also taught in Plymouth before becoming an upper servant in the household of the wealthy Lloyd family of Chester County, where, she later proudly recalled, as keeper of the keys to plate and linen, "everything passed through my hands." The Lloyds brought her from the servant's quarters into the parlor to converse with visiting ministers and friends. Elizabeth Collins remembered going "from house to house, to work at my trade." An account about Elizabeth Daniel described her as born a "poor, low, despised girl." Mehetable Jenkins, a well-known minister of the 1780s, was illiterate, but as one chronicler moralized, "From this class, persons eminent for their wisdom and piety, have frequently arisen."[21]

These spiritual success stories stand out, in the main, because they do deviate from the norm. Most ministers seem rather to have come from comfortable middle-class families where economic struggle did not accompany spiritual struggle. Of the early seventeenth-century ministers, two ran small businesses to support themselves, but almost 15 percent chose husbands who were or became Public Friends. Thus ministerial families were a significant minority in the early seventeenth century. By the late eighteenth century, a smaller percentage had ministerial husbands, but in line with the shifting gender base, more had female relatives in the ministry, especially sisters. After the early Society-building phase, the number of women attracted to Quakerism from both upper and lower classes may have declined as the ministry became more firmly lodged in the middle classes.

The descriptions surrounding the period that Brinton called the "divided self" are much more revealing than family and class background. Brinton did not find much evidence of this conflict in the women's accounts he consulted. Other journals do show that the "call" imposed particularly heavy burdens upon women, and a number of them spoke of their reluctance at first to accept the need to devote their lives to following the light. Hoskins recalled that she herself had "spoken much against women appearing in that manner [the ministry]." She continued to rebel against the call, even offering to sacrifice her natural life "to be excused from this service, but it was not accepted." After six or seven months of this inward struggle, she finally spoke a few words in meeting and offered this general advice in her journal: "Oh, saith my soul, may all who are called to this honourable work of the ministry, carefully guard against being actuated by a forward spirit which leads into a ministry that will neither edify the church, nor bring honor to our holy High Priest, Christ Jesus." Men as well as women had to perform such self-searching, but women more than men had to satisfy themselves that their preaching was absolutely necessary for the Society. Elizabeth Hudson in 1736 recalled her feeling of "unfitness for Such an aufule undertaking and fear of my being Misstaken respecting my being call'd theirto

& the Ill Consequence atending Such misstakes was Continually before my Eye."[22]

Some early conversions were helped along by "remarkable visitations," moments of intense collective spirituality that swept through the meetings, leaving an increased number of women who felt the obligation to become Public Friends. Ann Roberts recalled such years in Goshen in 1722 and 1724. Mary Emlen remembered one that occurred in Philadelphia in 1728. Susanna Morris reported another in 1732–34 when one hundred ministers came forth in the Philadelphia Meeting.[23]

As the number of women ministers increased in the society, there seems to have been some consciousness among them about their secondary place in the church government. American Quaker women achieved separate women's meeting at the yearly meeting in Philadelphia as well as the more frequent regional meetings by mid-eighteenth century. In 1753 and again in 1784 American delegations of women proposed at the London Yearly Meeting that the English do likewise. At the time that American women were pushing for more control in their own yearly meeting, there seems to have been a feeling of discontent with the male-dominated hierarchy. Explicit comments about feelings of discrimination within the church are hard to find, but at least one very clear statement remains. In 1751, Mary Weston, an English Quaker minister attended the women's monthly meeting at London Grove in Chester County. "I thought it a favour & high priviledge I had to sit amongst them," she wrote later, "for I was sensible the God of heaven eminently owned the Females in that Service; tho' so much despised & slighted as their Help in church Government is by the great & wise Men of the Age, even in our Society." The Deborahs, she wrote, would be "exalted in the power of divine Sone [Son] to go on conquering & to conquer till he came to reign."[24]

Deborah, the Biblical prophet and judge who led the men of Israel to march against the Canaanite captain Sisera, had already become the symbol of female ministry. Ministers soon began to use her model to recruit women into the ministry. Ann Cooper Whitall recalled that a male minister visiting in 1760 explicitly called women to follow the lead of Deborah by saying there would be Deborahs among them. Whitall wrote in her journal: "I hope thay ar on there way." The following year she repeated the hope in her journal with the query: "Shall we be brought out of bondage by a Debrow as the Children of Iesral was. o who is Debrow."[25]

To accept the call to represent God was in some measure to represent women in the structure, and that call still represented a sacrifice for women. As the search for the Deborahs became more widespread in the Society, women heard the call, but often reluctantly. A testimony for

Elizabeth Collins referred to a time in 1778 when "her mind became exercised under an apprehension, that she was called to bear public testimony." The expectation that women would not leave their homes could, of course, become an acceptable excuse for ignoring the call. Collins herself wrote to a friend in 1796 that "the enemy of all good, will endeavour to keep us back, by persuading us that there is no need of so much circumspection and care—that we can be as good at home." Collins urged that all "give up to the heavenly vision." Her own spirituality was such an important gift to her that she valued and felt compelled to exercise it. Although she sometimes "gave way to reason" and put it off, she would then hear spoken to her "inward ear" the threat of deprivation: "The language was awfuel; I was brought to see the deplorable situation I should be left in, if after receiving so precious a gift, I should neglect to improve it, and it should be taken away, and I left poor and distressed."[26]

Thus, women struggled to maintain their spirituality at the cost of both physical and psychic distress. Several spoke of expecting spiritual calm after the call, only to be faced with continual spiritual trials. "I was desirous," wrote Collins, "of putting this [little journey] off sometime longer, when his language was intelligibly communicated to my mind; the day is but short, thou hadst need to be industrious." Collins reported continual "stripping seasons," when she was misled into believing she did not have to speak in public. As Public Friends, these Quaker women found that their spiritual welfare forced them to maintain a public presence that was not easy in a world still urging their obligation to remain industrious at home and silent in public.[27]

There is also evidence, although sparse, of some women who were called but did not receive the acceptance of their meetings and of others who could never surmount the daily responsibilities that called them back to worldly cares and kept their souls off center. But both male and female ministers encouraged those women who, they felt, had a genuine call and would be a credit to the church. Jane Hoskins of Chester County received such support. A male minister told her employers, the Lloyds, to adopt her and let her go where truth led. Male preachers, she recalled, assured her the Lord dispensed bread, flesh, and wine to women as well as men, "from thence inferring the Lord's influencing females, as well as males, with Divine authority, to preach the Gospel to the nations." On the other hand, there are a few reports of women who were disqualified. The best known was Jemima Wilkinson, who subsequently established a religious community in upstate New York and adopted the title of Universal Friend. James Emlen reported that her first call occurred in a New England meeting where, after five Friends asked her to stop speaking, she refused to be silenced. Hannah Barnard was disqualified from her ministry in Hudson, New York, in 1802 for her interpretation of certain parts of the Bible. One Pennsylvania anecdote reported Priscilla Deaves, on the elder's orders, being carried yelling from a meetinghouse after she preached against a decision of the meeting. During the schism that separated Orthodox and Hicksite Quakers in the 1820s, some Philadelphia leaders denounced minister Priscilla Hunt. Hicksite Quakers supported her, however, and she spoke at several city meetings despite murmurings against "ministers from the Country and other distant places." Such unpleasantness seldom occurred, for few Quakers, except in time of factional crisis, pursued the call if the meeting disapproved.[28]

One other reason beside home or community for not heeding the call could be—books. Elizabeth Hudson, like other Quaker women, felt her first call quite young, before marriage, but did not enter the ministry until later. Hudson, however, described her conflict as being not with gaiety or frivolity, as others sometimes did, but with learning. Withdrawing to a friend's country estate, she indulged her love of books until, finally, after agonizing days, she gave herself up to religion. Her account is worth quoting at length:

> Our hearts became truely United to Each other I believe not Inferior to that degree of friendship wch Subsisted betwixt Jonothen & daved wch Strength of love Induced me to leave my fathers House and Spend most of my time wth her at their Country Seat wch was a Situation that Suted well my Inclination to retirement and had also the oppertunity of their Liberary in wch was a good Collection of Books that at times I Entertained my Self with having Some lust of Books and Indeed in time found had too high a relish for them they being very Ingrossing both of our time & thoughts and I finding this to be the Effect of my Studies, found it best to deney my Self of them wch was no easy task.[29]

But one should balance this painful decision by Hudson to quell her lust for books and the consequences for female learning with what must have been a much more common deterrent. Among the diary entries of Esther Collins for 1787 are the following: "Washing, not so inwardly thoughtful as could be desired. . . . Many cares & Cumbers too little time for Sollid retirement. . . . I have for some time had a desire to retain things in my mind, but it seems too much crowded and choked with other cares. . . . had fresh desires to lay hold on that which will endure. . . . brought home 50f of flax which is likely to take my attention for a time." A widow with six children had almost insurmountable problems in heeding the call. Inner voices and the visual "drawings on my mind" forced Quaker women early in American history to deal with a problem modern career women are still facing, how to manage a household with all its fragile psychic and physical dependencies while also living a public mobile life. Consistent with the tradition of deemphasizing personal affairs,

Quaker women's biographical and autobiographical accounts are casual in their references to children. In the early eighteenth century, only eight accounts mention children, some as casually as "several," a "few," or simply "step-children." Martha Chalkey had five who all died before the age of three. Elizabeth Webb had eight. The difficulties may have increased in the late eighteenth century, as more women ministers married younger and thus probably bore more children. Ruth Walmsey had seven children, Susanna Lightfoot had nine, Hannah Foster twelve, Susanna Morris thirteen.[30]

Women ministers suffered both from being unable to leave home because of children and from having to leave them. In 1790, Collins reported: "It was no small trial to leave home at this time, having several small children, but was favoured to get where I could leave them to the care of Him, who is the great care-taker of his people." Ruth Walmsey left a six-month infant to visit a meeting in 1788 but did not mention to whom she entrusted her child. Ann Shipley's mother-in-law apparently looked after her children. Mothers, mothers-in-law, female kin, and hired women took up the burden of child care for mothers who felt the call deeply enough to leave small children.[31]

A limited ministry seemed to have been the solution for many women. Rachel Barnard recorded "but little freedom to go on formal visits," apparently because of home responsibilities. Susanna Morris did not begin going abroad until after the twelfth of her thirteen children was born. She wrote in her memoirs that God had helped her husband to give her up "to the will of him who has Called." Elizabeth Hudson, in 1751, referred to a period of depression and then recovery, but of being dissuaded from a visit four years later "by some." One older ministerial couple reported both traveling after their son was old enough to take over affairs.[32]

Early journals record concern for both husband and children. Arranging for communications was a constant battle, and letters were an almost unimaginable relief when they brought news that all was well, to be followed, of course, by the immediate realization that delay in the mails might mean the well-being of a family could already have worsened even as letters announced good health. Fragile as the link was, the letters joined these absent wives and mothers attending to their "Father's business" to the families left behind.

If travel fragmented family ties, it also gave women new relationships. Brinton quotes a woman minister warning single women against developing interests in their male companions. Presumably, married women needed no warning when traveling with male companions. Nor were there warnings against female relationships. Often two women ministers traveled together, younger with older, in what seems like an apprentice role. Women developed deep attachments for their companions. Susannah Morris wrote of her companion Phebe Lancaster: "And we had Joy and Gladness In Each Others Company." Jane Hoskins later recalled of her travels with Elizabeth Levis: "We were true yoke-fellows; sympathizing with each other under the various exercises whether of body or mind, which we had to pass through. . . . I hope the love which subsisted between us when young, will remain to each other forever; mine is now as strong to her as then." Hoskins wrote, "I am persuaded, where companions in this solemn service are firmly united in the true bond of Christian fellowship, it must tend to confirm the authority of their message, testifying their joint consent to the doctrines they teach, to comfort, strengthen, and support each other, through the many trying dispensations, which in the course of their travels they have to wade through; this being the real case, judge how great must be the disappointment, when it happens otherwise!" Sometimes it did happen otherwise. Wilkinson reported that the contrast in abilities made one of her journeys with Hannah Harrison miserable. Harrison had so large a gift, "a full stream to Swim in," while she seemed in a "Land of Drought."[33]

The most complex network of feelings is revealed in an exchange of letters between Elizabeth Hudson and Hoskins during a visit to England in 1749. Hudson and Hoskins formed an attachment so strong that it threatened to interfere with their spiritual well-being. "The want of thy Company makes every thing Insipid," Hoskins wrote to Hudson early in June 1749 on their first separation for independent gospel work. Seven days later Elizabeth replied: "I hope my good Master dont require my doing nature Such violence as to Continue long from Thee, so that as soon as this Comes to hand pray writ and let me know Where I may meet wth thee." Once reunited, however, the magic moments ceased. Hoskins wrote to Hudson later, "[We] never after found our Spirits united in Gosple labour as had ever before Subsisted which made us truly Dear to each other & help meats indeed and I was in hopes it would have revived again Especiale when the friend [another woman minister] parted from us who I Thought the principle Impediment to our fellowship as formerly, but this I leave & only Say I had a Bitter Cup to Drink during her being with us, for altho twas with me as I have hinted I dearly loved my Compn & Could not properly bear the thoughts of parting personally." After that parting, Hoskins reported "a Strang Struggle betwixt my affection to thee & duty to my Master & for Some did not know which would get the Mastery," her mind "a tumult of Jarring Passions." Finally, the power of religion subdued her passion, and she went on with her work. Such mid-eighteenth-century evidence of passionate relationships between women were precursors to those which Carolyn Smith-Rosenberg described as quite frequent by the early nineteenth century.[34]

Although these experiences of sisterhood must have done much to lessen the tensions that women felt in their male-dominated occupation, they did feel the constraints of gender. Elizabeth Webb, at the peak of her powers, touring New England in 1698 while intolerance toward women preaching still existed, wrote: "had I been a man I thought I could have went into all corners of Ye land to declare of it for indeede it is ye Great day of new england's visitation." This extreme constraint soon disappeared with the general acceptance of Quaker women in New England, but women continued to struggle against external disapproval and their own socialization. Several ministers reported being harassed by preachers and young men in their ministerial work. The existence of even these infrequent comments indicate that women did recognize and write about the problems of their place in eighteenth-century society.[35]

Women shared with men the continuing problem of when and how often to speak. Richard Bauman has noted the contradiction between the Quaker emphasis on silence to achieve the light and the need of ministers to be alert to openings when speech was not only appropriate but necessary. Ministers agonized over their initial call to preach; they also agonized over the appropriate balance between speech and silence during the course of their ministry. According to Quaker tradition, ministers did not consider themselves personally responsible for their words because they were moved by the spirit to speak; therefore ministers praised or blamed the preaching of other ministers very seldom. Advice on the spoken ministry nonetheless existed. After women were admitted to the annual meetings of the ministers in 1718, they no doubt received general guidance in their roles. Male ministers often offered informal support and criticism as well. Wilkinson recorded a male friend who said she would be better "if I was to speak more deliberately," and another who said she should have "spiritualized a little on some of the last expressions." Women felt the strain of speaking in mixed-gender groups. Sarah Cresson was conscious of her own reluctance to speak, "a danger attends me of shrinking in the presence of Men." Mehetable Jenkins never learned to speak well in large meetings, although some valued her lack of polish as being more fitting than speaking well, as some other women did. Elizabeth Foulke confided to her journal in 1788: "I often think I am one of the most stammering speech of any that ever were sent forth on such an errand."[36]

Accounts reported the efficacy of the Quaker style when once perfected by women. Comments often appeared in journals that women "appeared lively in the Ministry," spoke in a "very lively manner," or delivered their comments in "innocency and brokenness of spirit." Women developed their own styles. Rebecca Jones was known for delivering short comments, sometimes a single sentence. Phebe Trimble from Goshen received praise for a few "clear, pertinent, comprehensive, and savoury" words. Ruth Richardson's communications were reported as being "delivered in weakness and fear, and in much trembling." A style of oral interpretation of the Bible that emphasized improvisation rather than exegesis allowed women without scholarship to develop great talent. Like men, women seldom depended on memory but upon the revelation of God that brought remembrance. Elizabeth Hudson once fasted "to keep my spirit free and lively." When women were visiting, outsiders sometimes came especially to hear them speak, something that Quaker ministers increasingly opposed. Even Quakers might be disappointed to attend a meeting conducted entirely in silence, as some women ministers reported.[37]

The subjects of these improvised sermons by women are rarely reported because ministers normally did not record their own preaching and even opposed others recording their words. A number of the parables and descriptions of the content do exist, however. The loss of, or failure to record, the sermons of the early Quaker women ministers is a great loss to the history of oral tradition as well as to women's history. No sermons of Lucretia Mott, probably the greatest Quaker speaker of the nineteenth century, were recorded for her first twenty years of ministry. Only a handful exist from the last thirty-five years of her work when she had already achieved national reputation as a speaker. Therefore, the themes and metaphors used in some of these early journals are of importance because they are fragments of that larger oral testimony that has been lost.[38]

Few complicated metaphors have been preserved from the preaching by these women. One is given here, as an example of what might have been a pattern by women, referring to particular parts of the Bible much as did black gospel songs in the nineteenth century. Elizabeth Hudson was reported in 1749 as beginning a meeting with these words: "And in that day Seven woman Shall take hold of the Skirts of one man Saying we will Eate our own bread & wear Our own apparel but will be called by Thy name to hide Our reproach." One hungers for an explanation of such a powerful, and yet obscure, parable.[39]

Religious metaphors often include alternative ways of referring to deities that may reflect changes in social relations. Women preachers used metaphors that changed over time in rather significant ways. One of the earliest ministers, Hoskins, referred to "the great Lord of the harvest." Elizabeth Wyatt described herself in the 1740s as "a sharp threshing instrument in the hand of the Lord." The image of the reaper did not reappear in later speaking, however. Instead it was replaced after the 1760s by images of a very different diety. In two accounts from the 1790s, he was referred to as "the Great helper of his people," "my Holy Helper," and as a Lord who would carry lambs in his arms and lead those with young up the mountain. By 1816, Elizabeth Collins would say, "We are all children of the one Great Parent." Not enough accounts exist to do more than indicate that the change in conceptualization of the deity from an authoritarian to a loving God seems to have been reflected in the choice of language, the change in symbol from distant and threatening—the reaper—to loving and close—the parent. The domestication of the deity has been noted in other American sects as well. It may be that the presence of so many women ministers in the Quaker church helped this transition to occur earlier than in other Protestant denominations.[40]

Almost simultaneously with the change in personification of the deity came a new attention to social habits in the writings of Quaker women. It is perhaps too much to say that these themes were not present in earlier sermons since so few have been preserved, but they do begin to appear in writing as more women entered the ministry in the late eighteenth century. In the 1760s, a few Quaker women ministers began to question the use of strong drink, the use of tobacco "in all its shapes," as one said. At the same time, there was more prominently expressed a concern for children who were not being properly influenced by male parents. Children shared the Quaker birthright of parents, informally from the early eighteenth century and formally through the establishment of residence requirement for charity after 1762. Therefore, Quakers did not emphasize conversion and redemption as did the early Puritans. They did not, however, lose concern about their children's spiritual welfare.[41]

Beginning in the 1760s, women ministers began to talk about the need for children to be present at meeting, and for parents to be watchful of their children's material as well as spiritual conduct. In one of the most elaborate statements, Wilkinson spoke directly to women in 1762, telling them to bring up children in the scriptures and to "nip vice in the very bud in their children." Children were to be educated in plain apparel and economic independence; mothers were not to indulge them. She offered mothers a parable of fencing in young trees with a hedge because they were growing near a busy highway and could be destroyed. About the same time, Ann Cooper Whitall recorded women speaking in meetings of their concern of children and of her own concern that her children would follow her husband out of the church. "So cold about religion & children men grow now," she wrote. Fathers, she believed, had changed their older control of children, replaced it with a "fond affection" that led to indulgence. Hurtful liberties, she called them. The old would have the sins of the young to answer for, she warned.[42]

The perceptions of Whitall may have been correct. Fathers, as they lost religion, may have also lost authoritarian attitudes that older religious attitudes had reinforced, and threatened to draw children away from the church. In this way, as patriarchal power declined, women may have had to assume the burden of a moral conscience for men and children. To accomplish this, women may have asked reinforcement from the church of a spiritual equality not to be found elsewhere in society, a new authority that held women there and paved the way for an increasing number of women preachers in the late eighteenth century. This concern with children probably reflected a withdrawal of increasing numbers of middling male Quakers from church participation before the Revolution. There is at least scattered evidence, literally fragments in women's journals, to indicate the perception of a growing difference in the social concerns of men and women and the feeling that women must assert their control over children. The appearance before the Revolution of the theme of women's need to assume responsibility for the religious education of youth suggests that the ideology of motherhood, absent in most formal writings of the time, may have been unfolding in the Society of Friends as it did somewhat later among evangelical women. Quaker women seem to have been among the first American women to assert their need to control children in order to inculcate values that men ignored as they became preoccupied with secular, nonfamilial concerns. At any rate, Whitall, a Quaker with a heightened religious and social concern, welcomed women preachers. She wrote of Elizabeth Daniels: "O how she did prech in our meeting, o that we may remember it."[43]

Women ministers, in turn, encouraged greater involvement of women in church affairs. The influence began young. The copy books of young Hannah Minshall in rural Chester County for 1795–96 show the use of women ministers as models of emulation for young women. In her copy book, in extraordinarily clear penmanship, are these "Verses on Elizabeth Ashbridge by M.W.":

A worthy Friend from Goshen went
To warn the people to Repent
People that liv'd beyond the Seas
For to shew forth her Makers praise
To shew what God had done for her
That she might call to Nations far
That she might call to far and Near
She left her Friends and Husband Dear
Through many Towns she Travelled
The gospel Truth that she might spread
Her Visits satisfaction gave
Which by Accounts we have Receiv'd
Soon was expected to the Shore
But we shall never see her more
For she is gone we hope to rest
To live forever with the Blest
She met with Trials very deep
But yet the Sense of God did keep
She to her Friends did often show

As if her Cup did Overflow
Did Overflow with the new Wine
That came from him that was Divine.[44]

While occupying what was by now a traditional place in the Quaker ministry, these women expanded that role and drew more women in turn into active participation. Elizabeth Drinker wrote in her diary that the 1803 yearly women's meeting in Philadelphia was so crowded with "zealous women"—their reported number was between sixteen hundred and seventeen hundred—that many could not get in. Yet men still retained a dominant position in the Society. For example, in 1807 in Centre, Pennsylvania men visited erring women members on occasion, but women never visited men. Still, women had become a massive, visible, public presence.[45]

This influence was not confined to Quakers. Quaker women ministers, because of their traveling ministry, spoke before large numbers of people outside the church. This occurred in two ways, either by seeking out particular groups to speak to, as early Quaker missionaries had done in Boston, or by non-Quakers attending meeting. It also, of course, occurred more informally, as when Ann Moore encountered the Presbyterian minister at the tavern. During the eighteenth century each of these types of encounters are documented in accounts by women. They include preaching to Native American Indians, Afro-Americans, prisoners, sailors, and laborers, as well as a wide variety of visitors to Quaker meetings. Because of these contacts, the example of women ministers was not confined to Quakers alone, so that declining proportions of Quakers in the population were not the sole way to measure the influence of women ministers.

Women began preaching to Native Americans in the 1700s. Susannah Lightfoot preached "with power," according to one Quaker biography. Yarnall recorded irately that upon her visit to the Tuscarora Indians in New York, she learned a male minister was trying to induce the men to stop women from speaking in their assemblies. Nevertheless, she spoke. Susannah Hatton visited the Delawares in 1761. Important women greeted her as "sister," joined at silent worship, and then listened thoughtfully to her preaching. It was, her companion wrote later, "the most Melting season I ever saw." In a larger meeting that day, her words were translated and her message given that the "inward Mentor" spoke to all colors, nations, and denominations. After the message, she went among them shaking hands. Ann Mifflin visited Cornplanter's Seneca town in 1803 and talked to the principal women there, receiving their official call for the help of women to teach them white women's skills. Hunting was declining and women were obliged to do all the fieldwork while men lacked sufficient new occupations, according to the Indian women. Two years later Rachel Coope, the first Quaker missionary, arrived to answer their call.[46]

Quaker women also spoke to groups of blacks and participated in the debate among early Quakers as to whether blacks could or should be converted. Elizabeth Webb had a vision, which she recorded, of working among blacks as well as whites. And in 1781, women of Birmingham requested the admission of the first black to the Quaker Society. They lobbied until Abigail Franks was finally admitted in 1784.[47]

Other women preached to white audiences up the Delaware River, in New Jersey, in Boston, and in Rhode Island as well as in the South. Although women refused to minister to "itching ears," or be moved by people who sought formal preaching or "discourse," they continued to visit and speak to persons outside the Society into the early nineteenth century. Margaret Judge, in the 1820s, had meetings for sailors and their families and visited the New York penitentiary where she spoke to prostitutes. Esther Elliott and Alice Wilson met with "the laboring people at the lime Quaries" in 1825. Thus it is wrong to consider the Pennsylvania women Quakers as having a decreasing influence in either Quaker society or the larger majority culture.[48]

The Grimké sisters, when they arrived in Philadelphia in 1832, reported the Quaker women to be conservative, but this should not be misinterpreted as a gap in the tradition of Quaker women as forceful, and often radical, public speakers. The Grimkés found themselves amid an Orthodox Quaker meeting influenced by an evangelical Protestantism that expanded women's role in the traditional religious sects but had nothing to offer women like Lucretia Mott or Ann Wilson. These ministers not only continued to expand women's influence within the society but also increasingly transferred that influence to still wider circles of women outside.[49]

Quaker women ministers thus provided a crucial role model for rural women, especially during the late eighteenth century. Circulating through the backcountry and speaking when rural women and young girls attended Philadelphia Yearly Women's Meeting, they offered a model of public activity available nowhere else. In fact, rural areas overwhelmingly resisted the influence in the urban Orthodox Quaker meetings that brought, among other things, a diminished role for women after 1829. Twenty years later, in 1851, the progressive Hicksites of Chester County held the first statewide women's rights conference after Lucretia Mott failed to move the Philadelphia women to action. The expansion of the role of Quaker ministers was surely one of the crucial parts of the history of the women's movement in America. It formed the foundation upon which some Quaker women built outward to provide role models for ever widening groups of women.

To risk the opprobrium of the wider public was no easy thing in the early nineteenth century despite a Quaker tradition that gave women a usable past for supporting new ventures. That past not only drew upon the social tradition offered to Quaker women but also offered a spiritual discipline that would give them methods to create the psychic strength to resist criticism from an almost unified public when they first began to gather people together to speak of women's rights. The spiritual discipline that led Elizabeth Collins in 1796 to write "Centre then, O my soul, more and more, within the enclosure of the walls of its salvation" could also provide inner strength to meet a world that did not share a vision of equality for women.

CHAPTER 10

"Not Only Ours But Others" Teaching Daughters

Given the example of mothers as ministers, it is not surprising that Quaker daughters were among the first young women to enter the public sphere as teachers. Teaching was the first middle-class occupation opened to women in the nineteenth century. Teaching also signaled women's emergence into full literacy and into the reform currents that eventually led to the women's rights movement. Thus, the ideas that marked the acceptance of women as teachers in the opening decades of the nineteenth century have attracted considerable attention from historians.

In their search for the ideological origins of the women's rights movement, historians have already defined two distinct ideas. The first Linda Kerber has labeled "the ideology of the Republican mother," a cluster of attitudes popular in the years after the American Revolution that emphasized the importance of mothers being educated in order to educate their children. These ideas, argued persuasively by both male and female writers in New England and the Mid-Atlantic, were set forth most eloquently by Judith Sargeant Murray, the New England essayist who stressed the need for female education in the 1790s. By the 1820s, a second ideology was taking form, one that I would like to label the ideology of the teaching daughters, in which writers argued the benefits of women teaching.

The transition to the second ideology is easily traced in New England writers, from Judith Murray in the 1790s through Charles Burroughs in 1819 to Catharine Beecher in 1829. Women and men formulated out of their Republican experience the ideology for educating women, but in their arguments there was little hint of the revolt that would carry their daughters out of the home and into a new public sphere as teachers in the next fifty years. Murray, for example, argued that women should be educated, but the course of studies she suggested was to be taught to girls at home by their mothers. When she stressed that young women should be able to earn a living if necessary, she used as her example a woman who did needlework for pay so that she could resist inappropriate suitors. By 1829, in contrast, Beecher saw teaching as the alternative to marriage, a "*profession*, offering influence, respectability and independence."[1]

Building on, rather than replacing, the earlier ideology of Republican mothers' duty to teach their children to be good citizens, the new prescription to teach the children of others provided the rationale for a new public place for young single women. With the exception of Quaker women ministers, women previously held few public positions of leadership. If still bounded by a sex sphere ideology, the new public space for women sanctioned by the ideology of the teaching daughters allowed some loosening of the social bonds of young women and provided an important link to the feminist movement that emerged in the 1840s.

A similar but not identical movement took place in the Mid-Atlantic states. In 1819, Emma Willard suggested to the governor of New York that women would have to be trained as teachers if the ideas of Republican motherhood were be carried into practice. Anne Firor Scott has argued that from the opening of her Troy Female Seminary in 1821, Willard intended to train women for teaching as well as for responsible motherhood. Similarly in the state of Pennsylvania, teacher training for women was well underway before the 1830s. When Pennsylvania began to charter schools in 1838, twenty-five female seminaries and academies applied for incorporation, and others existed that did not apply, like the ninety-six-year-old Moravian Seminary and College. Academies, seminaries, and boarding schools all provided teacher training for women, although the programs were never subsidized by the state in these early years as were those for males. Pennsylvania gave five colleges over forty-eight thousand dollars between 1830 and 1838 to train ninety-one male students to teach, but few of the men ever did. The superintendent of the common schools declared in 1837, "Nearly all turn their backs on the ill-paid and thankless drudgery, the first moment an opportunity offers." The state did offer female academies a subsidy after 1838, but most women paid for their own teacher education. A normal school for young women was finally established in 1848, and five years later, a third of the teachers in Pennsylvania were women.[2]

Between 1790 and 1850, Quakers pioneered in elementary and secondary education for girls and young women. Then Quaker teachers defected, first to the public primary schools when Pennsylvania established its educational system of common schools after 1830, and later to new colleges when they opened their doors to women after 1850. The period between 1790 and 1850, then, is an ideal one in which to examine the formulation of the ideology of the teaching daughters and the practice that lay behind it.

Rural Pennsylvania seems to have differed little from rural Massachusetts in terms of female literacy. In both areas, about 60 percent of the white females were still illiterate in 1775. By 1820, however, young native-born white females were probably nearly all literate, because the 1840 census shows almost complete literacy for females over twenty-one (see Appendix table 19 and figure 1).[3] But black female literacy lagged seriously behind, with only 50 percent of black women twenty-one or over literate by 1850, a percentage lowered by in-migration from southern states where black literacy was almost nonexistent because whites feared it would lead to revolt (see Appendix table 20). In the Philadelphia hinterland, nearly all girls gradually achieved literacy in the years after 1775, which was the main reason additional teachers were needed. Republican mothers, no matter how patriotic, could not teach their children if they could not themselves read and write.

Among Quakers there is little evidence of an ideology of Republican motherhood. Quakers, with a few notable exceptions, did not actively support the Revolution. Instead, they underwent a period of internal reform that emphasized a revival of the concepts of the inner light and separation from the world. The doctrines of separation and reform played important roles in influencing female education, for church ministers, especially the women, emphasized the duty of mothers to give their children a religious education. But this preceded rather than followed the American Revolution. Because Quakers adopted birthright church membership for children rather than baptism or conversion, the church charged parents with responsibility for education of their children, including reading and writing. Beginning in 1746, the Philadelphia Yearly Meeting began urging more careful education of all Quaker children and the establishment of schools for those poor Quaker parents who were unable to educate their own children.[4]

The wealth and literacy of early Quakers should not be overestimated. Some were both poor and illiterate. Even those mothers who had a basic knowledge of reading and writing had few rural schools to which they could send their children for "grammar" schooling. In the countryside, there was often no educated group of male teachers available to teach grammar to children once the mothers had taught them the basics. The Philadelphia Quaker schoolteacher Anthony Benezet designed a grammar in 1778 for rural mothers so they could teach themselves and then their children.[5]

Self-education could serve as a temporary expedient in the backcountry, but after the Revolution Quakers renewed their efforts to establish schools for themselves and for the poor. And out of this movement in the 1790s came the first open advocacy of young women becoming teachers. In 1794, women ministers took the initiative in promoting women as teachers of poor children and of female Quaker children. Twenty-three-year-old minister Sarah Cresson recorded in her diary after the Philadelphia meeting of that year that minister Deborah Darby "encourage[d] Young Women, to undertake the care of not only their own connections, Younger branches of their own families, but also poor children." In the same year, minister Ruth Walmsey proposed "a plan for improving the female character, by employing teachers of their own sex to cultivate the minds and improve the manners of female youth."[6] The early 1790s, then, was the threshhold of a new era for Quaker women in teaching.

But what was so new about the pronouncements of Darby and Walmsey? Quaker and Anglican schools, particularly in Philadelphia, had already employed women to teach. The tradition of the dame school, where widowed or older single women taught basic reading, reached back into the seventeenth century in Pennsylvania as elsewhere in the colonies. A late seventeenth-century Philadelphian explained in verse:

> Good women, who do very well
> Bring little ones to read and spell,
> Which fits them for their writing; and then
> Here's men to bring them to their pen,
> And to instruct and make them quick
> In all sorts of arithmetic.[7]

Philadelphia school records list schoolmistresses during these years who taught writing as well as reading, though usually to poor black and white students, and rural school records occasionally list a schoolmistress.[8] What, then, was new about the call for women to teach?

Most important, perhaps, is the fact that all the pre–1790 literature assumed that men would teach the more advanced elementary schools for the poor, the regular primary schools, and the secondary schools for Quakers. In Philadelphia in the 1760 and 1770s, women usually taught reading, spelling, and sewing to Afro-Americans, while males taught reading, writing, and arithmetic to older black boys and the secondary school for white girls. One observer mentioned some girls writing at a Philadelphia school for blacks in the late eighteenth century, and an 1800 list of the Pennsylvania Abolitionist Society noted several teenage girls being taught writing and arithmetic. Records indicate that schoolmistresses at the Quaker School for Black People did not regularly teach the black girls writing until 1811, however, and there is no record of their teaching arithmetic to girls at all. Occasionally, a mistress filled in while the overseers looked for a new male teacher. Although the overseers several times mentioned looking for a mistress who could teach the same subjects as the master, they refused to offer a woman more than a third to a half of the master's salary even while expecting her to teach sewing and knitting in addition to all the other subjects. Overseers would raise the salaries offered to males rather than raise them for a qualified mistress. In 1798, for example, when both a master and a mistress complained of low wages, the master's yearly wage was raised to $500, but the mistress received only an extra "gratuity" of $20 in addition to her old salary of $150.[9]

Laments about not being able to find schoolmasters, particularly Quaker schoolmasters, lace many of the accounts of struggling Quaker school committees before the 1790s. Nowhere is there a suggestion that women *should* teach either the poor or Quakers, even though in practice women were so employed. Moreover, the women almost never taught secondary schools for young girls either. Thus, the calls by Darby and Walmsey to young Quaker women to teach moved a practice sometimes engaged in out of necessity into a positive principle.

None of this explains the cause of the change, however. One reason may have been the increasingly late age at which young Quaker women were marrying as the eighteenth century progressed; this left a longer period between maturity and establishment of their own homes. At the same time after the Revolution, middle-class Quakers were experiencing a new affluence, which enabled them to purchase consumer goods that lessened the responsibilities of young women at home. The older ideology of usefulness that had prepared young women for the business of married life persisted, but without new opportunities becoming available to them for other types of work outside the home. That women ministers took the lead in suggesting the new occupation indicates they may have been particularly sensitive to the position of young middle-class women and, perhaps, worried about the spiritual consequences of the lack of work for them.

By the 1790s, as more young women moved into winter classes and remained in school longer, male schoolmasters may have had difficulty disciplining them. When Joseph Hawley, a young rural Chester County Quaker kept school in the winter of 1796, over 40 percent of his students were girls, some of whom seem to have been responsible for his giving up teaching after the first year. "I had an average of 17 Scholars," he wrote in his journal. "The Smaller ones were Unruly, and the large ones of the Feminine was hard to be governed by me, for they were of Equal age and acquaintance whose tender Sensibility of Amorous friendship could not be Exhausted, nor could I Refrain from showing a propensity for their Love and friendship."[10] Whether or not this particular problem drove other young Quaker men back to the fields, there were continual complaints about the difficulty of finding enough male teachers for the Quaker schools in the 1790s as more women attended them.

In the presence of such shortages, whatever the reason, two conclusions seemed logical. First, older girls needed some place where they could go to study, and second, having acquired a better education, they could teach at some of the schools where it was so difficult to maintain a male teacher. Because women had already taught some of the less affluent students, the transition would not be difficult. "Not only ours but others" seemed an easy slogan to put into practice. Not radical at all—or so it seemed.

As young women turned to teaching in the 1790s, older Quaker women also manifested a new interest in the schools. They began to serve on school committees and as representatives to joint committees, and their names appeared more frequently on subscription lists circulated to raise money for the school building boom of those years. In one subscriber list that survives from Providence Meeting in Chester County, for example, 50 percent of the people listed as donating between 1793 and 1795 were women, who contributed over 39 percent of the amount collected.[11] Quaker women also began to form associations to organize their teaching of the poor more systematically. The Philadelphia Quakers formed the Society for Free Instruction of Female Children in 1796; and the Wilmington Quaker women formed the Harmony Society in the early 1800s. Both groups hired women to teach.

Although Quakers were among the first women to enter teaching, there were others as well. Sunday schools provided a place for women of all major Protestant denominations to exercise new roles and occupy new space. Teaching, like education, could be a means of enforcing conformity to women's sphere, but it could also be a means of resisting the encroaching bonds of "true womanhood." The tension between literacy as a potentially liberating change and as a way to enforce cultural conformity was present for late eighteenth- and early nineteenth-century women as it was for every other group of newly educated people. For the moment, however, literacy, education, and teaching seemed to offer liberation.[12]

Esther Fussell was perhaps typical of the young Quaker women who first discovered teaching as a new occupation outside the home. Fussell, a Chester County woman by birth, lived across the border in Maryland where her family had moved to obtain cheaper land in the 1800s. The Quaker meeting there established Forest School where Fussell taught 111 students in the years between 1815 and 1818, before she married at thirty-six. Although uneducated in grammar, Fussell taught her young pupils reading and writing and took them for rambles to nearby creeks to study nature. She often quoted romantic poetry in letters to the man she would soon marry.

Robert Wells, in a study of several meetings, has estimated that the average age at marriage of those women born after 1786 was slightly over twenty-three years and that only about 45 percent of Quaker daughters married at all. If this was true of other meetings, it may have been the major impetus for the development of the teaching role. Marrying later than their mothers or not at all meant that many Quaker women would look to teaching as an occupation. And, in turn, as more women aspired to become teachers, the demand for secondary education also grew.[13]

The new education for young women aged twelve to eighteen attempted to blend the teaching of special female attainments with a more rigorous intellectual discipline that went beyond the earlier curriculum of reading, writing, spelling, and grammar or the acquiring of ornamental accomplishments. Elite schools like that of Madam Rivardi in Philadelphia gave young women such an education, which filled the new needs of the female citizen. Victorine du Pont, who used her training from the Rivardi school in managing the Brandywine Sunday School, never took a job as teacher and never paid her women teachers. But a half dozen other Rivardi graduates from families of declining fortunes opened boarding schools for young girls in the 1810s. Although the school did not train teachers as such, the less affluent graduates simply turned to teaching for a living.[14]

In about 1800, Quakers moved toward a more deliberate preparation of women for teaching, and soon boarding schools dotted the rural countryside. Flourishing into the early 1840s, they had almost all disappeared from the Philadelphia hinterland by the time of the 1850 census. With the exception of the church-supported Westtown Boarding School, which survives today, most of them existed only for the lifetime of a dedicated family.

Usually a family interested in education would establish their home as a boarding school, the husband acting as principal and the wife managing the living arrangements. Young women did most of the teaching, although men might teach special subjects like chemistry or astronomy. The schools usually had twenty to forty young women in residence. Westtown was again an exception. Here the girls and boys were supervised by a hired couple, and the school received important financial support from the Philadelphia Yearly Meeting. The other couples struggled with financing and never established lasting institutions like the women's seminaries of New England. Nevertheless, the Quaker boarding schools provided an important education for women in the Mid-Atlantic.

FIGURE 13. **Westtown Boarding School.** Reproduced with the permission of Friends Historical Library of Swarthmore College.

Like other female seminaries, they influenced their women students in three ways. They allowed them to experience collective living outside the home, so that as they chose friends from among their classmates, they formed permanent non-kin relationships that usually continued after school years through correspondence, visits, and sometimes mutual political activities. Second, they provided special teacher training and often teaching experience. Finally, they provided places where young female teachers could pursue further education with access to libraries, equipment, and tutoring from older teachers. Thus the influence of these boarding schools was enormous, even though most of them lasted only a few decades.

Westtown, Kimberton, and Sharon were the three most important of these schools in southeastern Pennsylvania. The first one, Westtown, was established in 1799 and was modeled after a boarding school opened by the London Yearly Meeting twenty years earlier. It was explicitly designed to bring rural Quaker women into the expanding network of educated women. Located in Chester County and admitting both young men and women, the school had become 70 percent female by 1824. The majority of females is explained partly by the rule that boys had to leave Westtown at fifteen, whereas there was no age limit for girls, and partly by the growing demand for more education for rural daughters in the face of a lack of college facilities for them. During the first five years of the school's operation, young women came to study from forty-eight communities in Pennsylvania, twenty in New Jersey, and fourteen in other states.[15]

Westtown emphasized plainness and academic subjects, not reform activities. It did, however, expect to train young women systematically and thoroughly and without the ornamental trappings that still occupied most upper-class schools. Westtown promised to teach girls "domestic employments" as well as useful learning—the first students were asked to bring a "pair of Scissors, Thread-case, Thimble, Work-bag and some plain sewing or knitting to begin with"—but the school soon acknowledged that many scholars did not wish to sew. Those that did, the "sewing scholars," were allowed to spend a third of their arithmetic class time in learning plain sewing, knitting, and darning.[16]

Young women in early Westtown did not receive equal treatment with the male scholars. Boys had the run of the woods; girls were restricted to the grounds. Boys had a tailor to mend their clothes; girls mended their own. On reaching fourteen, each young woman had to wear a starched white cap even while at play or be punished. There is a record of one young woman being made to write out verses in punishment for stuffing her cap into her pocket while she played. Such restrictions did not last long, however. By the 1810s, small revolts had removed the cap ordinance, although teachers still wore them.[17]

In form, the education seems to have been equal, although it was modeled on the traditional male secondary school of the time. Except for the fact that women knit and made samplers, they were expected to study the same subjects and to excel. Rebecca Budd, a scholar and later a teacher at the school, reported in her student journal that committee members visited to hear the students recite and to examine their copy books, and that visiting ministers exhorted the young people to be "scholars in the school of Christ." These early visiting ministers apparently conceived of the school as recruiting ground for future ministers, but for students it became something very different. For one thing, in 1802 the school inaugurated a teacher training program, and by 1824, 181 women had been prepared to teach.[18]

For another, the Westtown dormitories, with crowded beds and trunks in the third-floor chamber, permitted young women to experience a shared communal life with others of their own age. This experience drew them together into a sisterhood that they maintained after leaving school. Martha Sharpless, daughter of caretakers Joshua and Ann, wrote excitedly in her journal during the 1814 yearly meeting in Philadelphia of going home with Mary Budd, Martha West, Priscilla Kirk, and Ann Newbold, and of how much like Westtown it was to be surrounded by them. Rachel Painter, who entered the school in 1812, wrote later to her cousin, then also at Westtown, that she had chosen only a few close friends, but with those few she had formed a friendship she believed would be "only dissolvable by death."[19]

Westtown also provided a new place for older Quaker women in education, for they were involved in the governance of the school from the beginning. The Philadelphia Yearly Meeting appointed Rebecca Jones, a Quaker minister who had taught day school for decades in Philadelphia (as well as selling books and women's clothes in her parlor), and six other women to sit on the committee that organized the school. Although women were only 13 percent of the governing committee, their presence signified their new importance to education; nine more women had joined the committee by the time the school opened in 1799, bringing the figure to 29 percent. Half of the original teachers were also women.[20]

Kimberton School, which survived for only thirty-three years, was less famous than Westtown but one of the best known of the rural private boarding schools that flourished in the 1830s. Kimberton was almost entirely composed of students planning to become teachers. Although the school did not survive the death of Emmor Kimber in 1850, it provided during its existence a unique example of a school that not only trained young women to become teachers but also introduced them to the most radical reform movements of the day. Emmor and Susan Kimber ran the school and their daughters, Abby, Mary, and Martha, did the teaching, with Abby assuming the major responsibility for the curriculum after the 1820s.

Emmor Kimber, an early teacher at Westtown and a Quaker minister, and Susan Kimber bought a two-hundred acre farm in Chester Country in 1817 and set up their school originally under the name of the French Creek Boarding School for Girls. Advertisements from local newspapers of 1818 to 1824 describe the school as a large twenty-room farmhouse located about two miles northeast of the Yellow Springs resort on the road to Norristown, near a village with mills, a tavern, and tenements. The school offered reading, writing, arithmetic, bookkeeping, English, grammar, geography, composition, botany, painting, French—and, of course, needlework. Whereas Westtown had rather strict rules of conduct, the Kimbers promised no penal laws or rules, asked that the young women "come of their own consent," did not encourage parental visits, and gave students vacations in April and October. Brochures advised students to bring simple but substantial clothing that would not require frequent washing and a pair of good leather shoes. The school did not prosper at first. Emmor was in debtor's court in 1825 and offered the property for rent in 1827. But somehow the Kimbers managed to survive until Abby was able to assume greater responsibility for managing the school. By 1830 Kimberton was specializing in teacher training, and for two decades, it provided outstanding education for country girls who hoped to teach.[21]

FIGURE 14. **Kimberton School.** Reproduced with the permission of Friends Historical Library of Swarthmore College.

Over the years, the composition of the school gradually changed as the Kimbers became more preoccupied with the reform movements of their time. In 1824, a newspaper reported that of twenty-five students, 50 percent were from Philadelphia, many from adjacent areas, and the rest from the South and the West Indies. A roster from 1839–40 shows the composition had changed considerably. Of the fifty students then enrolled, only 20 percent were from Philadelphia, 32 percent from Chester County, and 30 percent from other rural Pennsylvania counties. Two students came from Baltimore, and none from any other area of the South or the West Indies. Esther Fussell Lewis, a teacher who sent her four daughters to Kimberton to receive the education she had lacked, wrote in 1839 that most of the scholars were "country girls," and that Kimberton was "a fine place to prepare teachers who all seem to have places prepared for them." Among the scholars were Elizabeth Mott, daughter of abolitionists Lucretia and James Mott, and Sidney Painter, sister to two prominent Delaware County reformers. By this time, the Lewis home was a station on the underground railroad, and Kimberton itself soon became a refuge for fugitive slaves. In 1840 Abby Kimber attended the first World's Anti-Slavery Convention in London with Lucretia Mott and other women delegates from Philadelphia. Kimberton students remembered in later years hearing about the refusal of the London conference to seat the American women and the attention they gained by their opposition to the exclusion. Life at Kimberton was certainly never dull.[22]

Student memoirs describe Emmor as a respected but at times tyrannical and overbearing man, who on meeting day might, as one student wrote, preach "something like a week." In the evenings, he scolded the young women for their rowdiness and noisiness. Susan, however, was warm and loving, and the three daughters smoothed away the difficulties between Emmor and the students. The school's schedule was rigorous: up at five, studying to six, breakfast, more study until nine, school until twelve, study until two, and school again until four. Abby became the presiding spirit, evolving a method of teaching that encouraged students to express their ideas and relate facts in their own language. After the hard discipline of learning, students would scatter to apple tree boughs or mossy rocks to study, wade in the creek, dive in the pool at the run, tramp through the woods botanizing, mount cherrying expeditions, or just enjoy the yards filled with roses, mock oranges, and honeysuckles. Abby, wrote Gertrude K. W. Thompson later, encouraged their love of knowledge and self-improvement, and insensibly they "absorbed a love of right and hatred of wrong and oppression which made us what was then called abolitionsists." Young women also accompanied Abby to temperance meetings, but there were, according to the letters of students, also evenings of singing and dancing, especially when Emmor was absent on business. One evening, wrote Pattie in December 1843, "we danced about an hour, and Miss Abby tried to teach us, but she knows just about as much about it as your black cat." Kimberton, and especially Abby Kimber, offered the country girls a special warm and loving environment, a place to learn and enjoy learning and an introduction to the most radical reform movements of the day. As such it was unique.[23]

Other exceptional boarding schools existed, each with its own special character. Sharon Female Seminary, established near Darby in 1837 by Rachel and John Jackson, a Quaker minister, was probably the most ambitious of the boarding schools but it lasted only twenty years, closing in the 1850s. By 1851, Sharon accommodated eighty young women who could combine teacher training with a liberal education emphasizing natural philosophy, chemistry, astronomy, physiology, geology, botany, and other branches of science. Sharon boasted over four thousand dollars worth of astronomical equipment, including an equatorial telescope, a sidereal clock, and a barometer, as well as a papier mâché model of the human system with removable parts for dissection, and a large collection of fossils and minerals. John Jackson lectured on various subjects and promoted learning through "familiar conversations" with students. Pupils lived as members of the family, with only such regulations "as good order could prescribe." They were exposed to Jackson's fascination with nature as the "living instructive truth" and his belief that one's intellectual

FIGURE 15. **Sharon Female Seminary.** Reproduced with the permission of Friends Historical Library of Swarthmore College.

nature was "as much a gift of God as the gift of grace, and we are responsible for the culture and improvement of one as for the other." Jackson's love of science and commitment to the use of appropriate teaching instruments probably made Sharon an outstanding educational institution. But student letters from the less prestigious Wilmington and West Chester boarding schools indicate that the young Quaker women at these schools too were being allowed to pursue their intellectual interests in exciting and new ways and to move out into teaching with a new purposeful freedom.[24]

Rachel Painter is an example of what teacher training could mean to the country girls of the 1810s. Rachel arrived at Westtown from East Bradford in Chester County in 1812 and left in 1815 after refusing an offer to remain as a teacher. In the last year of her studies, Painter had assumed some of the duties of a teacher—preparing 140 quill pens each day was one—and she wrote frankly to her cousin that she planned to accept a liberal offer to teach at a Quaker school in Alexandria, Virginia, rather than remain at Westtown where the supervisors expected too much work. Painter loved children and her new job, which she kept for over fifteen years because, as she once wrote to a relative, although she "could now do very well without it," she had such "a degree of infatuation in the pleasure it affords me that . . . I should part with it reluctantly."[25]

Painter's letters describing her job between 1817 and 1823 convey the delight of a young woman moving away from her family into a job that paid well, one she felt competent to perform. In early letters from Alexandria, which she called "a brisk little place for trade," Painter described her first teaching job and why she liked it, providing an extraordinary glimpse into the working life of an early teacher. She wrote to her cousin Minshall Painter on January 3, 1817:

> I have never regretted leaving home, tho at the time it was a very considerable trial and more so than any one but myself was sensible of. When I took into view the responsibility attach'd to my important station, in connection with the Idea of settling amongst strangers, a stranger myself to the ways of the world, I almost despaired of doing any good. But my intentions being good have been crown'd with success. I have endeavoured to act my part as respected my School in the best manner I was capable of, and from any thing that appears have given universal satisfaction. I have the pleasure of seeing my girls improve so as fully to answer my expectation, and as the saying is, fortune smiles favorably. I can keep myself genteelly and lay up 300 D. per annum, which brings me an interest of 8 per cent in bank Stock. My Committee (with a number of speculators) met here yesterday to have an examination of the girls in the different branches of learning and were much pleased with their performances many of whom expressed their approbation at the close of school.[26]

Three years later, when the school committee proposed a reduction in her pay of one hundred dollars a year, she wrote to Minshall:

> I very plainly told them I should not submit to having sufficient reason to believe that I should succeed better by taking the School upon my own footing as it is well establish'd and pretty generally known. I now take my chance for a share of the public patronage among 30 other fellow chips. I have 20 Scholars at 5D per qr. and 5 at 6D per qr. and have 4 new ones to enter the 1st of next mo. Thou mayst now Judge whether or not from my statement which is the best side of the bargain. I pay 6 Dolls per Mo rent for my School Room from the 1st of the 12th Mo. to the 1st of 4th mo. at which time the Mo. Meeting will furnish me with one rent free. So that I think my prospects are tolerablly favourable.[27]

In 1823, still eminently satisfied with her position, she wrote to Minshall: "I live much more easy than I could any other way, unless I set up for a lady. That would not add to my happiness, for I naturally possess an active mind which calls for full employment." Like a typical busy teacher, she concluded she would "like a bit more leisure."[28]

Rachel Painter was only one of the hundreds of Quaker women who had already begun teaching by the early 1830s when the states of Pennsylvania and Delaware established the first public school systems. By their commitment to teaching, they ensured that education for women would be expanded and accepted. Between 1775 and 1830, white women in these rural areas moved from less than 50 percent literacy to complete literacy and black women to 60 percent literacy. By 1850, almost 75 percent of white girls and 50 percent of black girls aged five to fifteen were in school in Chester County (see Appendix table 21). As early as 1842, women already held 32 percent of all teaching positions in public schools in Chester County and had incomes averaging 80 percent of the incomes of male teachers.[29]

Women did not fare so well where Quaker influence was less pronounced. In both Delaware County, Pennsylvania, and New Castle County, Delaware, women lagged further behind men in employment rate and pay. In New Castle County, black girls received little schooling at all in 1850, especially in rural areas, and there was growing opposition in the state to the education of any Afro-Americans. Some of the indentures of black girls in Delaware began in the 1830s to substitute a money payment to the girl in lieu of schooling. The brief statement on the indenture that it was "not expedient" to school the girl was a grim forecast that the painful gains in literacy would be difficult to extend to a slave state (see Appendix table 22).[30]

On the other hand, the gains in education for both black and white girls in southeastern rural Pennsylvania were impressive. The lives of two women from this area, Mary Ann Shadd and Ann Preston, may exemplify how the learning afforded to young women could become liberating literacy for the teaching daughters of the 1830s.

Mary Ann Shadd was born in Wilmington, Delaware, in 1823, the first child of Harriet and Abraham Shadd. Her mother was seventeen and from North Carolina; her father was a twenty-two-year-old shoemaker. Her grandmother sold coffee, cakes, and sausages on the street of Wilmington; her grandfather was also a shoemaker. By 1823, the year her grandfather died, her grandparents had accumulated property worth thirteen hundred dollars. During the next ten years, her father moved to a position of leadership in the Afro-American community, representing black Delawarians at a national convention of colored people in Philadelphia in 1830 and becoming an agent for William Lloyd Garrison's newspaper, the *Liberator*. He helped frame a protest against the current white movement to send blacks back to Africa as a way of solving the problem of slavery. In 1833, he was elected president of the National Convention for the Improvement of Free People of Color in the United States. That year Mary Ann, now ten, entered a West Chester Quaker school for blacks. The Shadd family soon joined Mary Ann in West Chester. Shadd worked actively in the underground railroad, sold shoes to Quakers, and prospered. By 1850s, the five sons and four daughters of the Shadd family were all in school, and the parents had accumulated an estate of five thousand dollars.

Mary Ann moved back to Wilmington in 1839 to teach, and during the next twelve years she taught at West Chester, at Norristown (just north of Kimberton), in New York City, and finally in Windsor, Canada. By this time, she was lecturing and writing in support of the emigration of blacks to western Canada, Mexico, and the West Indies. In 1854 she helped found the *Provincial Freeman*, became its editor, supported sending one of its reporters south with John Brown to Harper's Ferry, raised money for Brown's widow, and recruited black soldiers to fight in the Civil War. She moved to Washington, D.C., after the war and supported women's suffrage. At sixty, she received a bachelor's of law degree from Howard University.[31]

Ann Preston, the second woman, was born in Chester County in 1812. After training at a Quaker boarding school, she taught in the 1830s. She joined antislavery organizations, read Garrison's speeches, and in 1851 began to study medicine at the first women's medical college in Philadelphia. She spoke at the first Pennsylvania women's rights conference that year and spent a productive life as physician and women's rights activist.[32]

In 1851, Preston moved to Philadelphia after quitting teaching to begin medical training. She was ecstatic as she wrote her old schoolmate Hannah Darlington about her new life: "The joy of exploring a new field of knowledge, the rest from accustomed pursuits & cares, the stimulus of competition, the novelty of a new kind of life, all are mine." Only a year before, in the fervor of nineteenth-century romanticism, she had written to Hannah of her dreams for continual expansion and attainment of the "soul." "It is a glorious thought," she wrote, that "we may not sit down in dumb despair, we may read, think, discipline ourselves by study daily, & find our reward in an ever widening vision."[33]

In the lives of teaching daughters like these, literacy had truly become liberating. It had carried them beyond the confines of the home to a place in the public sphere, which they occupied proudly and from which they engaged in battles to extend the rights claimed by the men of the Revolution to new groups of Americans. On the crest of this new revolution, they could hardly imagine that some women might one day feel themselves trapped in teaching as earlier daughters had felt trapped in households. Teaching had given them an occupation and access to higher education. It had provided an essential transition—for some women at least—from a functional to a liberating literacy through which they could interact with the social and intellectual life of the new nation in ways that only males had done earlier. Some of the discontent of early nineteenth-century women surely came out of their perceived social and intellectual deprivation; but the confidence to express that discontent just as surely came from the large numbers of women who shared the experience of living and working outside the traditional household. The teaching daughters were an essential link between Republican motherhood and feminist sisterhood.

CHAPTER 11

"*True Earnest Workers*"
Reforming Sisters

On February 6, 1852, Hannah M. Darlington sat down in her Kennett farmhouse to write a letter to Jacob Painter, a wealthy Middletown acquaintance, about holding a women's rights conference. "The interest I feel in the cause impells me to it," she wrote, "yet in truth I scarce know what to say—if we could get up such a convention as we *ought* to have great good would result from it, but if we get up a *failure* it will be a misfortune to the cause, and tho I have unwaivering confidence in the eternity of truth and the certain advancement of great principles, that advancement may be retarded for a time by imprudent action on the part of its advocates. It is a considerable labor to get up a convention. This must be done by a few *true earnest* workers."[1]

With this discussion of ideology and practice, Darlington opened the planning phase of what was to be the first Pennsylvania conference on the rights of women. She was forty-two at the time and had no children. She and her husband Chandler shared their farmhouse with a female relative and her seven-year-old daughter. The Darlingtons' small farm was worth thirty-seven hundred dollars, according to the census taker who visited them two years earlier. Jacob Painter, the thirty-eight-year-old bachelor to whom she wrote, was the youngest son of Enos Painter, a farmer in neighboring Middletown Township, whose estate the census taker had valued at seventy-five thousand dollars in 1850. Although separated economically, Darlington and Painter shared a progressive Quaker persuasion, an interest in social reform, and a latter-day Enlightenment culture that joined them in a common concern about women's status in society.[2]

Since the original women's rights conference in Seneca Falls in 1848, Lucretia Mott had been trying without success to get the Philadelphia women to sponsor a state convention. Elizabeth Cady Stanton was not free to visit in the fall of 1848, cholera raged the following summer and fall, and even an eloquent defense of women's rights by Mott in December 1849 failed to rouse the Philadelphia women to organize. Mott attended the regional conference in Salem, Ohio, in April 1850 and the First National Women's Rights Conference in Worcester, Massachusetts, the following year. Lucretia Mott and twenty-nine men and women from Pennsylvania signed the call for the Worcester conference. Hannah Darlington was one of them.[3]

Even the exciting Worcester conference failed to move the Philadelphia reformers on the question of a women's conference. In fall 1851, Mott wrote to Stanton that only a few Hicksites were interested. Nowhere was the failure of urban women to support the fledgling feminist movement more clear than in Philadelphia. It remained for the rural families of Kennett and a few surrounding townships finally to plan the conference. Bayard Taylor later described these men and women of Kennett in his *Pastoral Sketches* as "zealous for temperance, peace, and the right of suffrage for women." But that was after the Civil War. Before the war, antislavery was more prominent than peace, women's rights in general more important than suffrage. But Kennett was the center and women of the little community could date their reform tradition from the early eighteenth-century New Warke meeting.[4]

Traditional histories of the nineteenth-century women's rights movement gave it a particular heritage. Suffragist authors linked it to the meeting of Lucretia Mott and Elizabeth Cady Stanton at the London antislavery conference, the exclusion of Mott and other Pennsylvania women elected as representatives from their antislavery societies, Mott and Stanton's subsequent meeting near Seneca Falls eight years later, and the calling of the first conference there in 1848. More recently, some historians have wondered about the eight-year lapse between the London and Seneca Falls meetings. Judy Wellman and Gerda Lerner have shown from their examination of petitions to the U.S. Congress that antislavery petition activity by women reached its peak in 1838, two years before the London meeting. Thereafter, petitions show a marked decline as a focus for women's political activity. So there is an unexplained gap not only in the activities of leaders Stanton and Mott but also in the activities of the masses of women who rallied to the antislavery standard.[5]

Peggy A. Rabkin recently argued in *Fathers to Daughters* that the base from which the feminist movement sprang was not antislavery at all but the interest of New York women in the married women's property rights law. She makes the point that fathers were increasingly interested in passing on family property to daughters as well as to sons and that the impetus for Seneca Falls was not moral but legal. The women's property rights movement, according to this argument, is the key to understanding the convention of 1848 because it undeniably preceded that event. The antislavery gap, in this interpretation, is closed by legal reform efforts.[6]

Even if this interpretation explained the women's movement in New York—and there are scholars who question the Rabkin thesis—it would not necessarily explain the motivation of the Pennsylvania movement. To understand it, one must examine the history of reform there, particularly the types of reform in which the people of Chester County engaged. Many documents remain from these struggles—petitions, memorials, tracts, and the rich collection of diaries and letters of Esther Lewis who lived in Vincent but was part of the wider circle of Kennett reformers. Temperance, antislavery, and women's rights were three great antebellum reforms, but the context within which each evolved was broader and more encompassing than any or all three.[7]

Chester County women had long been concerned about the effects of intemperance in the eighteenth century. As early as 1761, a tract by minister Elizabeth Levis, written at Kennett, laid out the arguments against the "too frequent use of spiritous liquors." Although Quakers John Churchman and John Woolman were also condemning the use of alcohol for harvest workers, there was a feeling in Chester County that this was a women's issue, for a friend had written to Levis, "What can we women do? The men uphold it." Levis wrote her tract in reply. The main occasion for excess seemed to Levis to be harvest time with its tradition of drinking in the fields. Although she did not mention how this affected women directly, she saw drinking as an occasion for both self-destruction and oppression of others. Masters, she argued, could no longer guide their workers in right conduct.[8]

If temperance surfaced in mid-eighteenth-century rural Chester County as a concern of at least some women, so too did concern about abolition, but only after women had helped introduce slavery into Pennsylvania. No antislavery tracts by women remain. What influence women had in the late eighteenth century seem to have been exerted within their monthly meetings between 1755 and 1774, when Chester County Monthly Meeting urged the Philadelphia Yearly Meeting to move from a policy of admonishing against owning slaves to one of disownment for selling or transferring slaves except to set them free. The statements by male Quakers during this period are eloquent in their determination to liberate both blacks and whites from the system. Wilmington minister David Ferris, for example, wrote in 1766 that "negro captives" must be delivered from their captivity and Quakers freed from slaveholding. Blacks, he said, were "fellow creatures." Although they left no ideological statements, women did participate actively in the Quaker manumission movement of the late eighteenth century. In 1806, Quaker minister Alice Jackson first proposed a boycott of the products of slave labor.[9]

During the first three decades of the nineteenth century, Chester County males moved into associations to work together toward abolition of slavery. Some twenty joined the Pennsylvania Abolition Society, and others formed the Chester County Abolition Society in 1820 to provide education for black children, establish a fund for refugees, and oppose the kidnapping of blacks. Most Chester County activists advocated legal reform and gradual abolition, but by the early 1830s a group of Hicksite families in the county were forming antislavery societies in support of immediate national abolition. The slavery issue did not move women into public until after the Hicksite split of 1829. The newly organized antislavery associations petitioned both state and federal governments for abolition, but women confined their efforts to the support they could offer from their homes. They did not ask to mingle in the groups publicly nor did they create their own groups.[10]

The 1830s changed all that. The movement to organize women into antislavery associations was not confined to Quakers or to Pennsylvania women, but in areas of Hicksite influence these women quickly assumed leadership in the growing movement. That leadership emerged from a broad-based mass movement that linked country and city together in Pennsylvania.

In that state, the growing black population was the most important influence in the movement. During the 1820s, freed and escaping blacks settled in Chester County in large numbers, especially in New Garden Township. Migration from Delaware was particularly heavy, with settlement across the state lines facilitated by the demand for black labor in the rural areas of Chester. By 1830, black communities existed in West Chester and Wilmington and in rural townships, but precariously (see Appendix table 23).

The formation of separate black churches was the first evidence of their presence. Few blacks followed Abigail Franks into the Quakers after the 1780s. From the comments of Quaker women ministers who conducted services for blacks in the eighteenth century, it is clear that the style of Quaker services—increasingly controlled and silent—was not popular with blacks who felt the spirit manifested itself in movement and sound. Most blacks first joined Methodist congregations, but as black communities grew in the late eighteenth century and expressed an interest in establishing separate black congregations, first Philadelphia and then Wilmington Methodist whites moved to spatially segregate the large minority of black parishioners in their churches. Segregation extended even to communion, the symbol of Christian unity. In protest, blacks withdrew and founded all-black congregations. These churches soon created their own powerful institutions, which developed a style of African Methodism suited to their spiritual and cultural needs. In 1814, the Wilmington congregation began a yearly meeting that came to be called "Big Quarterly" and drew both free and unfree blacks from Delaware and Pennsylvania into spiritual and cultural unity. In 1816, West Chester blacks formed a Bethel African Methodist Episcopal church there. By 1837, almost nine hundred black churches existed in Delaware and Pennsylvania to serve many thousands of members.[11]

Wilmington had the most members and the most hearers. Philadelphia had fewer, but New Garden and New Brittain townships and New Castle, Delaware City, and Christiana had sizable churches. These had become centers of anticolonization thought by 1820, and although blacks did not join Quaker meetings, they saw them as allies in their struggle for freedom. At the same time, in Wilmington and West Chester, a group of secular black leaders also began to emerge. They worked with Quaker ministers in organizing politically to oppose slavery ideologically and to help refugees escape from captivity and settle in southeastern rural Pennsylvania.[12]

Historian Larry Gara has argued that the emergence of this first black liberation movement was essential to the success of the underground railroad and rightly pointed out that only small groups of Quakers supported this movement. But Quaker assistance was not as unorganized as he indicates. In Chester County, the movement was especially well organized among rural Hicksites. The growing consciousness of blacks in northern Delaware and southern Pennsylvania of their need to oppose slavery, colonization, gradual emancipation, and the enforcement of the Fugitive Slave Act was a major element in the changing attitude of the more liberal Hicksite Quakers toward their role in that struggle. But the organized collaboration of that small group of rural Quakers in Pennsylvania and the large number of blacks they assisted in the struggle is well documented and real.[13]

Quaker activity depended on and drew upon the collaboration of a specific network of women and men to aid refugees as well as to change the climate of opinion among the Hicksite Quakers and within their neighborhoods. Most of these people did not believe during the 1830s and 1840s that electoral politics could change the institution of slavery. They did, however, believe that they had a mission to collaborate with the black free population in assisting refugees and to lobby in nonelectoral ways to change the sentiment of their communities. Quaker women in rural Pennsylvania played a crucial role in both these activities between 1830 and 1850.[14]

The underground railroad took form during the early 1830s, which were days of high mobility and immigration into Pennsylvania. By this time, blacks were asking for and obtaining help from Quakers in the effort to move refugees north in violation of federal fugitive slave law. After making contact with Thomas Garrett in Wilmington, refugees would go to Quaker farms across the border in Kennett, then up to northern Chester County, and on to Philadelphia or directly north to Canada. Garrett, who kept careful track of the numbers of assisted refugees, started fourteen hundred along their way before 1848. Esther Lewis kept no accounts, but her diary records taking "blacks to the railroad." Her tenant Maris Norris often carried them by wagon as he moved back and forth to the Philadelphia market with country produce.[15]

One of the best-known escapees assisted by Hicksites was Rachel Harris. Harris had worked at Kimberton School before marrying and settling with her husband in West Chester. Her earlier owner swore out a complaint, and West Chester officials took her into custody and incarcerated her at a judge's house, planning to return her south. The alert woman managed by a strategem (she probably got permission to go to the privy) to go out to the backyard of the place of her confinement at Church and Miner streets, where she scaled a seven-foot fence and made a dash for freedom down the back alleys to the home of a trusted woman Quaker friend. By evening, she was at Esther Lewis's home. Lewis wrote in indignation to her daughter: "She will stay here to day, and go further this evening. Poor wretch she is hunted like a partridge on a mountain. She has been a resident of W.C. five or six years, thought herself secure, and with her husband had acquired considerable property." Harris apparently left for Bucks County that evening dressed as a male and from there made her way to Canada.[16]

The work of clothing in northern dress, feeding, and nursing refugees went on at the Lewis home for over two decades. Graceanna Lewis later recalled, "There was never a time when our house was not a shelter for the escaping slave."[17]

The passage of a new, more stringent fugitive slave law in 1850, together with hardening resistance by the blacks in Chester and Lancaster counties led to open conflict. In 1851, when a group of blacks resisted reenslavement with arms in Christiana, just across the Chester County line in Lancaster, Quakers helped care for the dead and wounded southerners but also sent black resisters along the same network through Chester County and on to Canada. Black women who stayed behind later joined the men there. Thus the illegal underground activity in Chester County went on for twenty years.[18]

At first, it seemed as though the legal antislavery movement of the 1830s would follow the same course as the abolitionist movement of the 1820s. Males would organize public groups, petition for changes in state and federal laws limiting the civil rights of blacks, and attempt to change the attitudes among their white neighbors and in the country at large. In fact, the second phase of the nineteenth-century movement began in just this way. Dr. Bartholomew Fussell, Esther Lewis's brother, was one of the early signers of the Declaration of Sentiments of the first Philadelphia Anti-Slavery Society formed in 1833, but neither Lucretia Mott nor her other female relatives who attended this meeting signed the document. Instead, the women formed their own Philadelphia Female Anti-Slavery Society. The same year, local Chester County organizations began to take form. The Clarkson Anti-Slavery Society in the southwest and then societies in Bradford, East Fallowfield, Uwchlan, Willistown, and other townships formed, and in November 1837 these organizations sent delegates to organize a Chester County Anti-Slavery Society.[19]

From the beginning, these organizations encountered stiff resistance from the general public, so strong, in fact, that organizers found few allies even among the Hicksite Quakers. At a meeting in the Willistown schoolhouse, opponents threw a hornet's nest through an open window. At a meeting in 1837, a mob surrounded the courthouse where one group was holding a meeting and threw cayenne pepper into the stove and rotten eggs through the window. Local organizers brought suit against the leaders of the mob in an effort to halt the violence and to defend their right to assemble peaceably.

The Lewis daughters attended that 1837 meeting, but they were still a part of the small minority of women who believed it was acceptable to attend mixed meetings. The sentiment of those few had been fueled by the writings of Elizabeth Chandler, a Quaker born just across the border on a Centre, Delaware, farm. Chandler had not taken part in earlier organizing, but through her antislavery writings of 1825 to 1834, she helped change women's perceptions of what they could do publicly to bring an end to slavery. She had only written, but in those writings she urged women to action.[20]

Blanche Hersch has called attention to the extraordinary role played by Chandler, a little-known writer who began publishing her antislavery work at nineteen. She stressed in her romantic rhetoric that women had a responsibility to their black sisters in bondage. In her early poems like "The Kneeling Slave," she exhorted her readers to "Pity the negro, lady! her's is not, / like thine, a blessed and most happy lot / Thou, shelter'd 'neath a parent's tireless care, / . . . But her—the outcast of a frowning fate." Later she wrote "Letters on Slavery: To the Ladies of Baltimore," where she urged women to unite to carry emancipation plans into effect. Chandler continually sounded the theme of responsibility, association, and immediate action. After Benjamin Lundy published her collected works in 1836, Esther Lewis and other backcountry Hicksites bought her books in lots and circulated them to neighbors and friends. Soon after publication of Chandler's works, stationery began to appear in Chester County, bearing the picture of a shackled black woman and the logo "Am I not a Woman and a Sister," the same that had appeared in

Garrison's *Liberator* in 1832.[21] Chester women did not take an active part in the first Philadelphia Female Anti-Slavery Society in 1833, but by 1837, they were distributing antislavery tracts, arranging for male advocates to speak in local churches, and campaigning to win the support of neighbors to the antislavery cause.[22]

Some sense of the new political awareness of the young countrywomen is conveyed in letters of Rebecca Lewis, a daughter of Esther Lewis, who moved to Kennett to teach in late 1837. She attended the meeting broken up by the mob with cayenne pepper and rotten eggs and tried to shield the male speakers by sitting in the window, assuming the mob would not attack her. She was right; they did not. Rebecca had as her bedfellow in Kennett a good abolitionist woman who belonged to the Anti-Slavery Society and subscribed to Garrison's *Liberator*. Rebecca wrote to Esther, "It does me so much good to have *one* think just as I do." Later they were the only two women to attend a public meeting to discuss what could be done to counter mob action against antislavery advocates in other parts of the country. Quakers at Kennett seemed to be reluctant to take action after meeting, however. Some even refused to open the meetinghouse to abolitionist lecturers. "It makes me feel very unpleasant to be in such a place—and feel there are those around me who have not hearts to feel for the oppressed in our land," Rebecca wrote.[23]

In the Lewis family, both mother and daughters worked to spread the ideology of abolitionism. While Rebecca taught antislavery attitudes in her classes and attended public meetings, Esther preached more quietly at home. She related stories of outrages to visiting friends, giving them copies of Theodore Weld's *Slavery As It Is*. She wrote after one such visitor had left that she had not continued the conversation too long so as not to pour "water too fast into a narrow necked bottle." Thus conversation went on at home and abroad.[24]

Out of this grass-roots activity came the petition movement. This movement began early in 1836 with the publication by the Philadelphia Anti-Slavery Society of an address to the women of Pennsylvania, asking their support for a petition campaign. Eventually, thirty-three hundred women of Philadelphia and its vicinity signed petitions. Undoubtedly, they were circulated at the Philadelphia Yearly Meeting, for the name of a Quaker woman headed the first petition, and the signatures of Lucretia Mott and Sarah Grimké both appear on it. The women asked that slavery be abolished in the District of Columbia and the territories and that the slave trade between the states be ended.[25]

Women took the movement back to their communities, where they carried petitions from neighbor to neighbor. After the May 1837 women's antislavery convention in New York City, the women's campaign increased in the hinterland. During that fall and winter, thousands more women signed their names to petitions, asking Congress to abolish the slave trade in the District of Columbia, among the states, and in territories. Philadelphia women climaxed the campaign with a long petition bearing almost five thousand names early in 1838.[26]

This massive campaign was the first time in American history that women collaborated to exercise one of their political rights as citizens, the right to petition. The insecurity about their political status is evident from the variety of ways in which they addressed the petitions. The longest ones were printed memorials by "undersigned Female citizens of Pennsylvania, and parts adjacent." Smaller hand-written petitions showed important variations in how women described themselves. Some used the title "ladies" and the word "inhabitants" instead of "citizens." One Chester County petition proclaimed its signers "women." A Westmoreland petition had two neatly divided columns with "single ladies" in one, "married ladies" in the other. An Erie County petition with both men's and women's names had three sections, one male "voters," a second male "citizens," and a third "ladies." A petition from Allegheny bore a number of names in the same handwriting with Xs after them, apparently indicating that the women who were not able to write signed with a mark. The petition campaign did not penetrate far into the illiterate ranks of society, however, for almost all the women not only signed but signed with a firm clear hand. The row on row of names is eloquent testimony to a people newly awakened to their political rights.[27]

Antislavery petitions reflect a geographical pattern of abolitionist sentiment in the hinterland. Out of thirty-three counties submitting petitions, Chester, Washington, and Allegheny counties sent in those with the largest number of women's names. Most petitions were gender segregated, with more males than females signing. When petitions were mixed, more males than females signed, and the names were usually segregated. On only a few were the names mixed together.[28]

This outpouring of female political action was greeted in Congress by a move to restrict the newly used political right by refusing to consider the petitions. Historians are still debating other results of the campaign. Many of the petitions were destroyed, which makes the evidence uncertain, but as both Judy Wellman and Gerda Lerner point out, there seems to have been a drastic decline in petitioning activity at the national level by women after 1840. Petitions from Pennsylvania in 1854 protesting the repeal of the Missouri Compromise are almost entirely male.[29]

What happened to the women? Lerner suggests that in Massachusetts the antislavery campaign shifted to a state civil rights campaign. Because Congress excluded abolition petitions from 1836 to 1844, it seemed more practical to petition the state. In Pennsylvania, violence intensified the crisis in the antislavery ranks, leading many of the more conservative urban Quaker women and men to abandon public activities. In May

1838, four thousand women and men met in the new Pennsylvania Hall to hold the second annual women's antislavery convention. A mob of ten thousand with another ten thousand spectators thronged the streets around the hall, burned it to the ground, and attacked a black church and an orphanage.[30]

Chester County women attended that spectacular meeting. Later they circulated reports about the burning in periodicals and by word of mouth. They did not cease their activity but continued to circulate antislavery petitions through August 1838. By the following year, however, women had difficulty in finding a place to hold their annual conference in Philadelphia. A study of women's antislavery activity in Rochester, New York, by Nancy Hewitt points to a defection of evangelical women from the antislavery ranks in the 1840s because of dissension over women's role in the movement. In Pennsylvania, a similar crisis occurred. When Chester radicals wanted to hold an abolitionist meeting in West Chester in May 1842, not a single church or public building was open to them, so they resolved to hold the meeting in the street.[31]

Although rural lectures continued, the emphasis shifted to state legislation for civil rights. In 1848, for example, hundreds of Chester County women signed petitions asking for state enfranchisement of colored citizens. Only 290 women signed federal antislavery petitions that year, and long petitions from Delaware County carried only men's names. Chester County activists continued to gather signatures of both male and females and the radical Hicksites continued to sign. A memorial from the antislavery society of eastern Pennyslvania bore the names of Lucretia Mott, Mary Grew, and Sarah Pugh. But the petitions seem to have represented only a hard core of committed female activists. These were the women who were emerging as the first Pennsylvania feminists.[32]

The progress of women within antislavery societies and the decline of those societies is reflected in the minutes of one group organized in northeast Chester County. Organized first in early 1837 as the Schuylkill Township Anti-Slavery Society, members changed the name to the Franklin Anti-Slavery Society in 1839, and then merged with the nearby Kimberton Society in 1840 to form the Lundy Union Auxiliary to the Chester County Anti-Slavery Society. It met in schoolhouses and in Friends, Presbyterian, and Baptist meetinghouses, for the last time in June 1847. Its fate seems to have been typical of many township antislavery groups formed by Hicksite Quakers in the late 1830s.[33]

From the beginning, the new organization admitted women to membership. Although no total membership roster survives, one has the impression that women soon outnumbered men as members. Lists of new members show that almost twice as many women as men joined during the first years. Esther Lewis and her daughter Graceanna, as well as Abby Kimbert, belonged. A resolution signed by members in 1840 contained the names of seventeen females and thirteen males. Men held all the offices at first and represented the group at general county conventions. In April 1838, however, the group authorized all women members to attend the women's conference to be held in Philadelphia in May (the one that was burned out by the mob), and in August of that year one woman was elected secretary-treasurer and another to the executive committee. Committees usually comprised nearly equal numbers of men and women from that time on. In October, the members nominated seven women and eight men to represent them at a state antislavery conference, and in November, when the male officers did not show up, Sarah Coates chaired the meeting and signed the minutes as president. Members elected a new male president to the faltering leadership at the next meeting, but when the organization revived and changed its name early in 1840, members elected Benjamin Fussell, brother of Esther and a strong women's rights advocate, as president. Sarah Coates then became vice president. Women thereafter appear in minutes as active discussants, proposers of resolutions, and even as the majority of delegates to a Philadelphia conference in May 1840.

During its decade of existence, the society reflected the grass-roots antislavery movement and the issues that gave birth to the organization and eventually caused its death. The political philosophy, as recorded in its discussions, reflected a deep commitment to Enlightenment thought. In discussing antislavery issues, members asserted their belief in the right to petition as being an inalienable right that Congress could not take away, however much it wished to avoid the controversy engendered by antislavery petitions. Members also talked of the American Revolution being unfinished "until we pass through a blodless & peaceful revolution" and all people including blacks were free. They opposed the use of force to free blacks but also opposed any assistance in reenslaving refugees. The right of black people to assemble peaceably and participate in political groups became an issue when the head of the local natural history museum refused to allow the county antislavery group to meet there because it included representatives from black antislavery groups. The society opposed a state law penalizing racial intermarriage as increasing "a feeling which gives rise to injustice & an interference with individual freedom." The society upheld the right of persons regardless of sex or race to belong to the American Anti-Slavery Association and opposed both colonization and paid emancipation. It denounced paid emancipation in the British West Indies as a great burden imposed by the wealthy on the laborers of Britain. The group continued through its short life to oppose organizing a separate antislavery party and urged male voters to oppose candidates who did not support human rights. It gradually moved toward supporting for Congress men who promised not to "give votes for slaveholders." The last entry in July 1847 indicated there were few remaining members and recorded the feeling that it was "not thought necessary to organize." Politics at the polls had replaced associationalism. Women were now isolated from men.

Dissension within Chester County Quaker meetings on antislavery questions continued through the 1840s. Progressive Hicksites were forcibly ejected from the Kennett Meeting for their antislavery stance. A Quaker wrote in 1845 that one member in ejecting another who refused to be silenced "gave his pantaloons a kick just as he was going out the door." Supporters refused to attend further Kennett Meetings and in East Fallowfield another fifteen resigned. Mobs continued to harass those who persisted in their antislavery activity. Electoral politics seemed not to be a major issue because antislavery militants had little chance of election to office and, in fact, opposed the formation of a separate anti-slavery party. Throughout the 1840s, those candidates who did run as abolitionists received few votes in Chester County. The formation of the Liberty party brought little change. As late as 1852, an abolitionist candidate won only 338 of 1,100 Chester County votes. Fusion of abolitionists with a major party would not take place until the late 1850s. Then, of course, it would be part of the monumental shift that fueled the election of Abraham Lincoln to the presidency in 1860.[34]

The most militant antislavery advocates of the Hicksite Quakers were not in a majority either in meeting or in their communities through the 1840s. Although their work continued, it was not popular. As the progressive Hicksites moved out ahead of public opinion on antislavery and as members quarreled within as to whether or not to form a political party, support dwindled. If political activity were to be the direction of the antislavery advocates, as it seemed increasingly to be, then women in their present disfranchised state would be unable to follow. They needed enfranchisement, but it could not be obtained on the basis of antislavery advocacy. Women needed a more popular cause.

Pennsylvania women did not fill the gap left by the declining antislavery movement with activity on behalf of the married women's property rights as did reformers in New York. The movement to obtain legislation guaranteeing property rights for women previously excluded from holding property after marriage flourished in New York during the 1840s. In Pennsylvania, however, this movement was not widespread, although Mott, Grew, and a few other women did gather signatures. Ten petitions, signed only with the names of men, were sent to Harrisburg in 1848, the year the legislature considered this most important act. The movement for married women's property rights did not involve women at the grass roots as had the antislavery movement. Women's political consciousness was not expanded by this legal battle but by the revival of an older concern—temperance.[35]

After Elizabeth Levis published her temperance tract in 1761, public discussion in Chester County merged into a general concern over harvest intemperance. According to one oral tradition, the Gibbons family of Westtown refused one summer to furnish rum and found no workers. Neighbors then pitched in to help with the harvest and decided not to provide rum to their workers either. This incident apparently occurred sometime in the late eighteenth century. Marietta cites Joshua Evans as also refusing to keep his agricultural workers in rum and paying them higher wages instead. The reform measure of providing nonalcoholic beverages to farm workers was probably complete among Quakers by the early nineteenth century, before the public phase of the campaign opened. Tavern licenses were issued in Chester County without much concern and certainly without public complaint by women until the 1810s. Women's names begin to appear on petitions in 1818 against issuance of tavern licenses. The first temperance societies were formed in Pennsylvania soon after, but not until the 1830s and 1840s did Hicksite women and men begin to move their concerns about intemperance outside their society.[36]

Males formed the first temperance societies. In 1832, for example, the Union Temperance Society in Chester County reported fifteen to twenty farmers had discontinued the use of ardent spirits in their agricultural concerns. Drinking, proclaimed the officers, led to murder, pauperism, and crime. By the 1830s, then, the workers were being transformed into a modern agricultural work force, expected to remain sober on the job. At least, the employers were establishing a policy of not providing liquor to their employees.[37]

There was no highly visible public temperance activity by women before the 1840s, when the abolitionist and antislavery movements had moved women first to public protest and then to dissension over the public forms of that protest. In Rochester, New York, public advocacy of temperance reform by women emerged immediately following the withdrawal of evangelical women from the public antislavery campaign. These groups now saw intemperance as the primary source of all vice, so threatening to society that women must publicly organize against it. Between 1840 and 1850, hundreds of rural women in western New York took up the new cause, forming societies and joining in petition campaigns. Chester County women followed a similar pattern of involvement in temperance. In 1842, Esther Lewis noted that in her neighborhood 130 people signed pledges not to drink within a few weeks. Large meetings were held in the Vincent area the following year. By 1846, there was enough state-wide support for temperance to enable the legislature to pass a local option law. Temperance crusaders mobilized in hundreds of small rural towns throughout Pennsylvania to pass local ordinances prohibiting the sale of liquor. In Chester County, the law passed by a large plurality. Then, two years later, the state supreme court held the law unconstitutional, invalidating the local laws in operation under its provisions.[38]

The antitemperance backlash in Pennsylvania was typical of trends in other states. In New York, temperance advocates like Susan B. Anthony were remarking about the "retrograde march of the temperance cause," as state legislatures began to respond to antitemperance pressure. The response of temperance women everywhere was a new militance and political activism. The emerging feminist leaders had found a grass-roots cause that touched large numbers of white women as slavery never had.[39]

The outburst that followed resulted in hundreds of petitions inundating the Pennsylvania capital asking for state prohibition. Almost one hundred petitions from eighteen counties remain in the legislative files in Harrisburg. Chester, along with Allegheny, Washington, and Delaware counties, sent over two-thirds of the petitions, with women being the most active in Cambria and Chester counties (see Appendix table 24). Over 150 Chester County women signed. In addition, they organized a public meeting at Kennett in February 1848 under the leadership of Hicksite women and then carried a long memorial to the legislature at Harrisburg recording their support for prohibition, which Sidney Pierce read before a legislative committee.[40]

In the memorial, the women wrote consciously as women "debarred from voting in public matters," as citizens who bore the consequences of the system that licensed the sale of liquor. While couched in terms of the "sanctuary of our homes," and men made "maniacs and monsters," the argument also cited practical reasons, such as secular law, natural law, and political rights. Intemperance was expensive, and it caused taxation without protection, it abolished the natural right of self-defense, and it violated the political rights of the majority who voted for prohibition. "Had our sex been permitted to vote on this particular question," the memorial concluded, "the majority would have been greatly increased." The address was sent to Harrisburg proudly as an address from a "Meeting of Women" in the name of womanhood.[41]

In the months following, Hannah M. Darlington, Sidney Pierce, and Ann Preston went from place to place in Chester County spreading the message. Hannah's husband Chandler drove them around. Hannah later recalled in a letter to Elizabeth Cady Stanton, "We addressed many large audiences, some in the day-time and some in the evening; scattered appeals and tracts, and collected names to petitions asking for a law against licensing liquor-stands."[42]

In December 1848, the women held an even larger temperance convention at Marlborough Friends Meeting Hall, this time addressing the men and women of Chester County. At this meeting the ideology of sisterhood became more pronounced, for the address to women began "Dear Sisters." The supreme court had held the local option law unconstitutional and the women began by asking "What must now be done?" The answer to this rhetorical question was to call women to the cause. "By meeting together and taking counsel one with another," the call urged, "we will become more alive to our duty in relation to this momentous subject." Here was a clear call to one gender with a separate interest: "If *men* will remain comparatively supine we must the more energetically sound the alarm, and point them to the danger." Although the women also made a call to the men, to "fathers and brothers," they saw a special role for women.[43]

Again the women appointed committees to obtain petitions in their respective neighborhoods and to call meetings to read the addresses. Women met twice more, in February and December of 1849. These meetings provide a vital link to the later women's rights convention, for the women active in the later convention were also active in the temperance meetings, speaking, organizing, and collecting petitions.

Much has been written of temperance as a response to immigration, one aimed at social control. Immigration of the Irish into southeastern Pennslyvania may have accelerated support, but Irish working-class immigration into that area was not new, and had been fairly constant during the early century. It is possible that middle-class men became more active during the early nineteenth century with the goal of social control. In the 1854 Pennsylvania plebiscite, Philadelphia and Delaware County males voted heavily for prohibition. But so too did males in western rural areas, the very areas that had staged the Whiskey Rebellion in the 1790s in support of free distilling.[44]

Rural women might have been responding to increased migration, but their public pronouncements and private letters do not single out Irish immigrants. Rather they explain their concern in the gender-specific terms of physical abuse of women and the disruption of their lives by male intemperance. Battering of women seems to occur most frequently where women are isolated from relatives and social institutions, and rural women for centuries had to deal with physical abuse as best they could, often without the support of local officials. Esther Lewis referred in her diary several times to husbands beating wives. In an era when the physical abuse of women was seldom discussed publicly, temperance rhetoric may have been a proxy for that discussion. At any rate, the consciousness of prohibition as a women's issue was an important link in the chain of reform moving women from private to public politics in the 1840s.[45]

Consciousness of women's rights had, however, formed part of the motivation for women's involvement in both the antislavery and the temperance drives. That consciousness began to surface in the Lewis correspondence in 1837. When a lyceum lecturer advertised in December of that year that women and boys would be admitted for half price, a group of young women at Kennett protested. The lecturer then reduced the price for men so that it would be equal for all. Private jokes went back and forth in the Lewis correspondence that indicate family conversations on the subject. Women's rights, joked Edwin, the husband of Rebecca Lewis Fussell, "are just what is Convenient to *men* you know!" To which Rebecca added: "If *you* do *know* that women's rights are just what is convenient to men *I* have yet to learn it." Graceanna Lewis reported with interest in 1838 an address by Abby Kelly in New England. After ministers had interrupted her defense of women's rights, wrote Graceanna, Kelly had finished with a "short but pithy and pertinent speech."[46]

Yet such private consciousness and support did not result in public meetings such as those held earlier by women for antislavery and temperance. The 1850 Worcester Convention in Massachusetts became a link joining Chester County organizers, and especially Hannah Darlington, to the emerging women's rights movement. She signed her name along with Lucretia Mott and twenty-eight other men and women from Pennsylvania for the call to the first national convention. She also attended the conference along with representatives from Chester and Delaware counties. Jacob Painter was one of the seven delegates from Delaware County. It was through the network of people who had attended the conference that Hannah heard Jacob might be interested in working on a conference and wrote of the need for "true earnest workers."[47]

In her first letter, Darlington sketched her idea for a women's rights conference for the eastern part of Pennsylvania, to be held for two days in late May or early June in the new Horticultural Hall in West Chester. She planned that a "few certain friends" would issue a call; a committee of correspondence would write inviting people to attend and would select officers and "drill" them to act with "propriety and grace." She stressed the solidarity of the sexes to elevate both men and women: "they must rise and fall together." It was clear from this first letter that Darlington had already contacted a number of women.[48]

In the next two weeks, Darlington contacted many more about the possibility of the conference. "Everybody says *have one*," she reported to Painter. The Kennett supporters and Lucretia Mott soon approved the time and place. All this preliminary organizing was done, as Darlington later wrote, "helter skelter, almost anytime & almost any how." As to rights to be discussed, Darlington listed legal, educational, and industrial rights. Social relations, she concluded, "come more under the *spheres* and are more matters of compromise than positive rights." She wanted to keep women's rights distinct from antislavery, to keep it "womanly," but strong of purpose and upon the "leveling up principle."[49]

The call, issued in April, reflected the influence of Darlington. Addressed to the "friends of Justice and Equal Rights," forty-seven people signed the call to discuss the position of woman in society, her natural rights, and her relative duties. As Darlington had advised, the call stressed the need to discuss ways to remove legal, educational, and vocational disabilities and proclaimed: "The Elevation of Woman is the Elevation of the Human Race." It did not mention the spheres.[50]

Both city papers reported the resolutions of the conference, including the slogan, "Equality before the law, without distinction of sex." They reprinted resolutions proclaiming that natural rights entitled women to an equal part with men in political institutions, that participation in government did not involve "sacrifice of refinement or sensibility of true womanhood," and that women should be represented in government. One West Chester paper attacked the men who attended the conference as "old women in pantaloons" and the women as old maids, amazons, and infidels. Still there was ample publicity for women who did not attend to learn about what had been advocated.[51]

The conference passed fourteen resolutions. Of these, a number were concerned with arguing what equal rights would *not* do—they would not sacrifice "the refinement and sensibilities of true womanhood," they did not reflect "conflicting interests between the sexes," they would not obliterate "distinctive traits of female character" but would make more apparent the "sensibilities and graces which are considered its peculiar charm," and they would leave the question of spheres, "the differences in the male and female contributions to take care of themselves." Two more resolutions simply expressed satisfaction at recent changes—that the exposure of the "wrongs of women" was being met by a kind spirit and redress of these wrongs, that the passage of the married women's property rights bill was evidence of the equity of their demands.[52]

Seven other resolutions asserted more positive demands. The first was that anyone claiming to represent humanity, civilization, or progress must subscribe to equality before the law, that women must be allowed to study the physical, mental, and moral sciences with men, that tax-supported colleges and universities should not exclude women, that women should inherit the property of the husband exactly as he did hers (in effect, a call for community property), that women should also have custody of children if they were qualified, and that women should be paid equally for equal work. The last two resolutions called for direct action on two specific demands—a political campaign to change the state laws relating to voting and inheritance rights, and an educational campaign to inform women of the rights and privileges accorded to women in other states. Despite the defensive and conciliatory tone of the resolutions, they covered four major areas—work, education, politics, and family law—and called for greater freedom for women in all aspects of their lives. The resolutions would serve as an agenda for over seventy years.

Of the letters read at the convention and the speeches delivered, it is noticeable that none mentioned antislavery. Although the *History of*

Woman Suffrage later argued that antislavery activities "were the imitative steps to organized public action and the Woman Suffrage Movement" and we have seen the importance of the earlier movements, the conference of 1851 did not mention this reform. A letter from William H. and Mary Johnson, read at the convention, referred explicitly only to "those benevolent associations particularly for promoting temperance, in which the females of Chester County have borne such a conspicuous and effective part." This was particularly true, they reiterated, when male associations in the state had languished.[53]

One theme throughout the years of antislavery and temperance movements and the emergence of the women's rights movement had been the increasing emphasis on egalitarianism. This was evidenced in appeals to "sisters," to sibling equality rather than authority. The sibling tie was a central image in each movement and provided a persistent theme in the 1830s and 1840s: "Am I not a woman and a sister," was an antislavery motto, "sisters" were called to the temperance standard, and "sisters" were not to be dependent on "brothers" but to achieve an equal status with them through the women's rights movement. With this theme came a rejection of hierarchical and patriarchal authority. Those who espoused this egalitarianism were a small minority, but they left the practice, and some theory, of an alternative to a patriarchal basis for society.[54]

The theme crops up repeatedly in letters and statements. If women were vague on womanhood, female character, sexual differences, and the desire for harmony, they spoke clearly on their denunciation of the doctrine of inequality. In a sense, the final speeches and resolutions reflected that careful balance that Darlington was determined to retain.

The resolutions upheld the possibility of "true womanhood" surviving women's entrance into politics. They said if women's province was to soothe passions no better place existed for that than in politics, that her "true sphere" was the one her nature and capability allowed her to fill and not "that appointed by man and bonded by his ideas of property"; and they dismissed the possibility of conflicting interests between the sexes. But they also offered special support for women in medical organizations and schools, and argued that property of "joint industry and economy" should go to the wife at a husband's death. With women's cooperation, predicted one speaker, women would be free in less than ten years.[55]

This overoptimism was consistent with the strong belief in men's and women's natural rights, but resolutions were worded carefully to deal with the commonly held ideology of spheres, true womanhood, soothing passions, and cooperation of the sexes. Speeches emphasized the needs both for married women's rights and single women's opportunities. The claims were moderate and yet, for their time, politically progressive. They were a fitting culmination of the reform movement of the radical Hicksite Quakers.

Knowing what these reformers did and said, however, does not explain their motivation. Who were these people and why did they participate in a publicly unpopular reform, becoming a minority that advocated changes that would take so long to achieve in society? The names of seventy men and women appear in the documents of the women's rights conferences of 1851 and 1852, and some biographical material exists for over two-thirds of them. Detailed census data is available for thirty of the rural participants who lived in twenty-four households, and these necessarily will be the sample for analyzing these politically active females. A sample of Kennett Township, West Chester, and Kennett farm households were compared with the twenty-four households to see if the activists were representative of the Kennett community and of the rural town of West Chester. How did they compare in ethnic, economic, occupational, and family composition?

The most important difference to emerge is ethnic. The women's rights reformers reflected the ethnic monoculture of the Kennett farms rather than the growing number of black and Irish households in the township of Kennett and in West Chester. The leaders could have invited a black woman to speak—Sojourner Truth spoke when the progressive Quakers convened later that year at nearby Longwood. Mary Shadd might have welcomed an opportunity to speak in her hometown. Even the middle-class Forten sisters who attended a later Philadelphia conference were apparently not present. Black women as well as Irish may have attended the conference—no accounts have yet been found that describe the audience—but they were not officially represented. Nor did Darlington ever mention concern about ethnic or cultural representation. In fact, her desire to separate women's rights firmly from antislavery may have led her consciously to avoid asking women of color to participate (see Appendix table 25).[56]

The figures indicate that the households from which the leaders came were relatively wealthy. Their wealth was far higher than the mean for West Chester, Kennett Township, or Kennett farms. Although individual households, like the Darlingtons, may have had modest wealth, overall they had double that of either township or West Chester households. Over 50 percent had female live-in help compared to only about a third for township and town. These were not families, then, who were losing out in their communities. A few professionals in West Chester—lawyers, one doctor, and a school administrator—had impressively large incomes. Thus one cannot argue, as some historians of reform have argued, that activism was the result of a general "status deprivation" of these families. It is possible that accumulation of wealth in Pennsylvania or other industrializing areas was bothering these reformers, for these rural households were enlightened enough to be concerned about economic differences.[57]

It is more likely, however, that status deprivation of females within these families was a factor in their activism. The families exercising leadership had growing numbers of aging daughters in their households. The difference in mean age of the children in women's rights households and other households is large. These daughters and their families may have been apprehensive about the prospect of being unable to provide an income through marriage alliances. They would be concerned, then, with achieving flexible educational, vocational, and legal rights. Without an expansion of their options, the daughters could have suffered status deprivation in relation to their own cultural background—being dependent and a financial drain on their families.[58]

The West Chester women's rights conference did not concern itself with the problems of women workers, as did the Rochester conference of 1849, which reported on the low wages of sewing women. A large percentage of women in Chester County were working outside their own homes, primarily as household workers, some as sewing women or proprietors of small shops in West Chester. Half of the reformers had at least one of these women working in their own homes. But the reformers seemed more concerned about their own work than that of others. Being a lady or a domestic, not a mill girl, were the main alternatives for these middle-class women. The radical Quakers rejected the role of the lady, but they also rejected the role of the domestic. They were not attractive alternatives for young middle-class rural women who chose not to marry. For that reason, Ann Preston, who delivered one of the main addresses, symbolized the possibilities for women in the profession of medicine.

Preston was exactly the type of young rural woman these reformers were thinking about when they spoke of educational, vocational, and political disabilities suffered by women. She came from a large Quaker Hicksite farm family of comfortable means. She had taught school and participated in the antislavery and temperance movements, but she could find no way to provide herself with a suitable income. There was no question that her brother would take over the family farm and provide for her aging parents, who were in their sixties. Other brothers had already left home to fend for themselves, and an only sister had died ten years earlier. At thirty-nine Ann Preston had just been able to embark on a professional career after the establishment of the Pennsylvania Medical School for Women. In a letter from Philadelphia shortly after the conference, she wrote that she did not want to depend on her brother.[59]

Of course, other women faced the same problems quietly, saw no violation of rights, did become dependent upon their brothers, and withdrew within their spheres. Without the progressive Hicksite strain of reform, nourished by the late Enlightenment as well as selected romantic ideas reinforcing an older egalitarian spiritual base, these women could not have become the first feminists to challenge their destiny as dependent women in a proscribed sphere. The loosening of the bonds did not come easily, despite family support, economic security, and an egalitarian heritage. What the women did was neither accepted nor accommodated outside their narrow reform circles. Rioters had burned down the Philadelphia hall where women had congregated to discuss antislavery. Philadelphia medical students had rioted when their sphere was threatened by women who wished to study medicine. Although the reformers came from a small unrepresentative group, they could have survived only within families and communities that nurtured their revolt.

That revolt and the extent to which it loosened the bonds of rural women should not be overemphasized. They did not challenge many roles. As Darlington had said, they considered social relations not a right but subject to compromise. They were concerned primarily with public institutions, and even there they chose their battleground carefully.

The issues of race, working women's conditions, and domestic relations were all subjects that would have to be addressed by later women's rights reformers. That they were omitted from the first struggle may have delayed the ability of white middle-class feminists to deal effectively with these issues in the next hundred years. Yet for all people struggling to loosen society's bonds, there was much to learn from these early feminists. The slow progress over the next hundred years was surely the responsibility of those who accepted their bonds rather than continuing the liberation process that had begun.

EPILOGUE

In 1850, the rural women of the Brandywine Valley must have been conscious of growing differences with women of the village of West Chester, two miles west, as well as the town of Wilmington, the "upstart village" at the mouth of the Brandywine and Christiana rivers that had grown to almost three thousand buildings, and the city of Philadelphia twenty miles east. Urban women in these towns and cities lived their lives amid an increasing concentration of male professionals—clergy, physicians, attorneys, bankers, public officials, and educators. Created primarily by males, the public spaces of urban areas were also mostly occupied by them. Both women and ethnic minorities found themselves relegated to particular spatial as well as social spheres, as the new male urban elite became more visible. Towns became more democratic for males—Wilmington men, for example, elected their first mayor in 1850—but they became less so for women. Even the women who identified themselves as part of the social sphere of the wealthiest men seemed to become invisible as the new urban male elite took form.

There were important exceptions. Some upper-class women joined middle-class urban women in voluntary associations to establish public space for themselves. Wilmington Quaker women, for example, formed their first Female Benevolent Society in 1800. As the early minutes show, women were to "arrange domestic affairs as to attend to the business of the Society in their turn, that each may take an equal Share." For the first twenty years, women gave out flax to be spun by poor women; later they furnished clothing, food, and wood. Women managed welfare contributions in money and kind from the city, male organizations, and individuals as well as from their own activities. In 1834, Quaker women also formed the Female African School Society to educate black girls. Women of more conservative religious denominations, meanwhile, formed the Female Colonization Society to support the sending of freed blacks to Africa. By 1850, Wilmington also had a Temperance Benevolent Society. Thus, middle-class women created and expanded a sphere that gave them an essential role in urban society. Upon these foundations, later nineteenth-century urban women established flourishing female organizational networks.[1]

Urban middle-class women also built a new female occupational structure. By the 1850s, even the village of West Chester had a female work sphere based on needlework, saleswork, and teaching. This occupational trilogy gave middle-class women a way to earn a good living, which also allowed them some status in the community. Wilmington had a similar job structure. Ann Rowan, known for her fine work, ran a tailoring establishment where she took in apprentices to learn the trade. Jane Alderice, a milliner who kept a dry goods store, according to one account "grew comparatively wealthy." When Dunn and Bradstreet investigated Mary Dixon's credit rating, they found she operated a small but profitable business with her unmarried daughters, and reported her to be a first-rate businesswoman.[2]

Working-class women, meanwhile, developed a culture and social structure of their own. Poor black and white women occupied a narrow occupational niche competently. They took in laundry, cleaned the homes of the middling and wealthy women, and trained their children in family survival. The black women of Wilmington's Chicken Alley, like Priscilla Durham, learned how to get help from middle-class women to survive. In West Chester, a growing number of Irish, like Susan Harley, took in boarders while their brothers worked as laborers.[3]

By the 1850s, the reality of urban life was different from rural life. The records of the Wilmington Board of Health document increasing concern over pigpens and full privys in the 1840s, prostitutes appear in the census of 1850 (the first municipal ordinance against houses of prostitution would be passed within a decade), and young women textile workers at the Franklin Manufacturing Company struck in 1852 for shorter working hours. Thus even smaller towns like Wilmington had become urbanized and industrialized by mid-century. The rural women who sold their produce and shopped for manufactured items on the streets nearby touched worlds far different from those they knew on the Brandywine farmsteads, despite the enormous changes that had taken place in their own world during the previous century.[4] They had no institutionalized volunteer groups, nor did they have the occupational opportunities of urban women, aside from teaching. Without the concentrated urban population in a time of slow transportation, rural women could not develop gender-based businesses, and craft occupations remained closed to them. A countrywoman who managed to enter the medical profession usually developed an urban gender-based practice.

Commodity production, which remained the main alternative for rural women, was based on the family farm. But the farm, embedded in patriarchal legal restrictions, offered women little ultimate control over wealth, property, or even their labor. Still, within these constraints, rural women experimented with and developed gender-based commodity production. They preserved, almost intact, a self-sufficient household economy that provided a firm foundation for the market production of the farms. They provided most of the uninstitutionalized welfare for the rural population and organized their own labor into efficient farm units, both as managers and as a work force. Women also developed a consumer economy that diverted a considerable amount of rural earning into the purchase of urban manufactured goods. Although some migrated to the cities, particularly widows and single women, most countrywomen found satisfaction and comfort in their rural lives. They also developed a self-conscious discontent that was manifested most visibly in the middle-class families who supported social reform. There was no widespread migration of rural poor to cities because rural people developed fairly sufficient economic structures for the majority and helped those who could not achieve success move west, where they re-created the old farm family with its dual productive model for use and market. Dependent rural poor did increase by mid-century, a problem compounded by the lack of concern in the larger urban-based communities. These dependent poor women were subject to a welfare structure that did not recognize their needs.

The modified rural patriarchal family of this Mid-Atlantic area was thus able to exist through the first half of the nineteenth century. Women had loosened the bonds of the traditional rural family in small but significant ways; it was reduced in size, less subject to male control, and modified by female experimentation and productivity. In some instances, women exercised political leadership. The rural economy survived through two major structural transitions, first in the 1760s and again in the 1840s. In the history of economic development, that must surely stand as a great accomplishment.

APPENDIX

MAP 1. Philadelphia Butter Belt, 1850

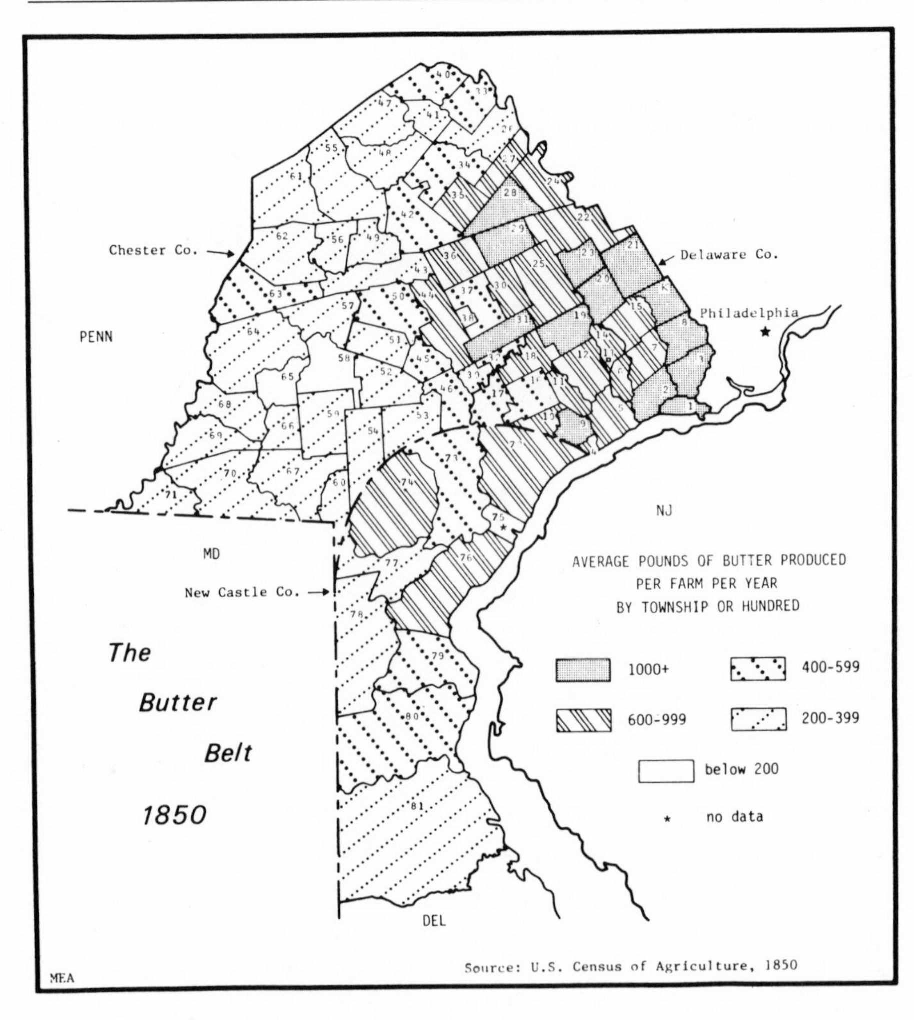

This map is based on a map created by Marley Amstutz.

Key to MAP 1, *opposite*

DELAWARE COUNTY
—Townships

1. Tinicum
2. Ridley
3. Darby
4. Lower Chichester
5. Chester
6. Nether Providence
7. Springfield
8. Upper Darby
9. Upper Chichester
10. Bethel
11. Aston
12. Middletown
13. Media
14. Upper Providence
15. Marple
K. Haverford
16. Concord
17. Birmingham
18. Thornbury
19. Edgmont
20. Newtown
21. Radnor

CHESTER COUNTY
—Townships

22. Tredyffrin
23. Easttown
24. Schuylkill
25. Willistown
26. East Vincent
27. East Pikeland
28. Charlestown
29. East Whiteland
30. East Goshen
31. Westtown
32. Thornbury
33. East Coventry
34. West Vincent
35. West Pikeland
36. West Whiteland
37. West Goshen
38. West Chester
39. Birmingham
40. North Coventry
41. South Coventry
42. Uwchlan
43. East Caln
44. East Bradford
45. Pocopson
46. Pennsbury
47. Warwick
48. East Nantmeal
49. East Brandywine
50. West Bradford
51. Newlin
52. East Marlborough
53. Kennett
54. New Garden
55. West Nantmeal
56. West Brandywine
57. East Fallowfield
58. West Marlborough
59. London Grove
60. London Britain
61. Honeybrook
62. West Caln
63. Sadsbury
64. West Fallowfield
65. Londonderry
66. Penn
67. New London
68. Upper Oxford
69. Lower Oxford
70. East Nottingham
71. West Nottingham

NEW CASTLE COUNTY
—Hundreds

72. Brandywine
73. Christiana
74. Mill Creek
75. Wilmington
76. Newcastle
77. White Clay Creek
78. Pencader
79. Red Lion
80. St. Georges
81. Appoquinimink

FIGURE 1. Signature-Mark Analysis of Chester County Wills, 1729–1850

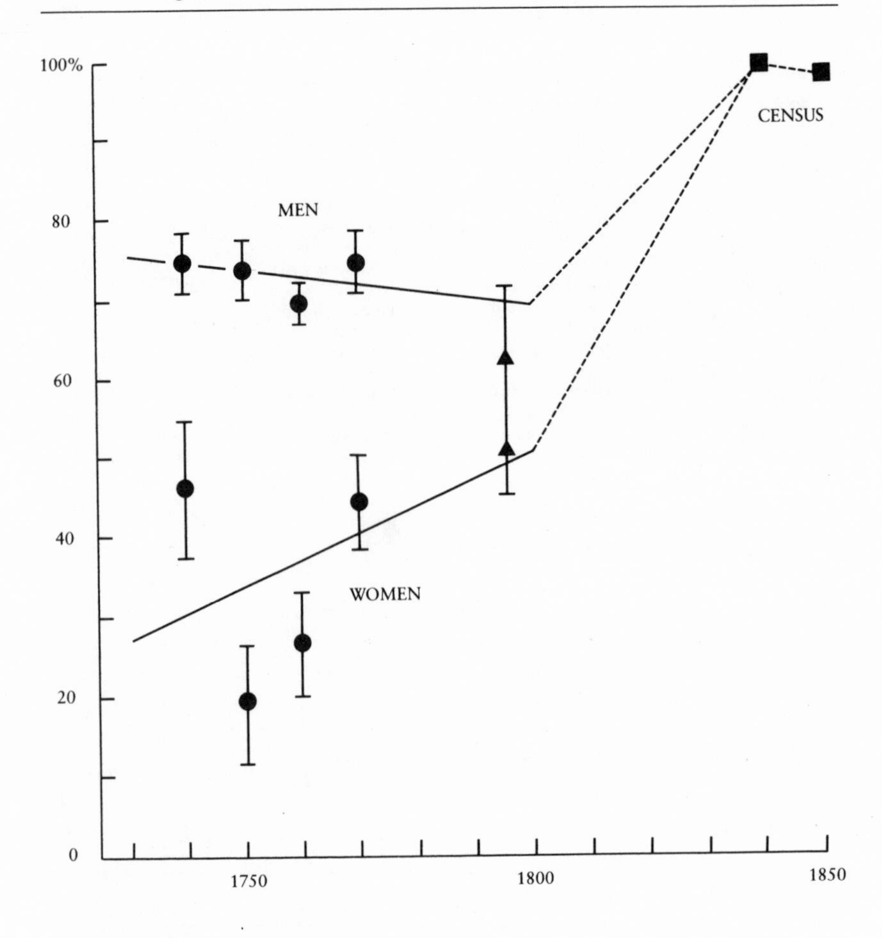

Data from Tully
Data from Chester County Wills, 1790s
Implied from census literacy data

Note: Vertical flags are estimated one standard deviation uncertainties due to sample size. Solid lines are best-fit linear regressions weighted for sample size.

Source: Chester County Courthouse, West Chester, Pennsylvania, and Alan Tully, "Literacy Levels and Educational Development in Rural Pennsylvania, 1729–1775," *Pennsylvania History* 39 (1972): 301–312.

TABLE 1. Chester County Religious Denominations: Number of Churches and Meetings

	1758	1800	1829
Friends	14	20	29
Presbyterians	8	8	11
Baptists	4	4	8
Episcopal	4	4	5
Reformed	2	2	3
Mennonite	1	3	3
Lutheran	1	1	1
Methodist	0	6	17[a]
Catholic	0	1	1

[a]Six of these met in private homes.

Source: Irene G. Shur, "Emergence of the Free Common Schools in Chester County, Pennsylvania, 1834–1874" (Ph.D. diss., University of Pennsylvania, 1976), p. 258.

TABLE 2. New Warke Families, 1715–1750

		N
Age of women at marriage	22.9	36
Age of men at marriage	25.7	42
Number of children	6.9	61
Birth interval in months	29.5	57
Age at start of childbearing	23.8	29
Age at end of childbearing	39.1	29
Number of childbearing years	14.9	56
Length of marriages	31.2	39
Age of women at death	70.6	27
Age of men at death	63.6	38

Source: Friends Historical Library, Swarthmore College, Swarthmore, Pennsylvania.

TABLE 3. Seasonality of Marriage and Conception

	Marriages (N = 52) (1720–1750)				Conceptions (N = 330) (1720–1775)			
	New Warke		Wilmington		New Warke		Wilmington	
Month	Index[a]	N	Index	N	Index	N	Index	N
1.	46	2	66	5	102	28	110	14
2.	69	3	119	9	131	36	63	8
3.	162	7	92	7	131	36	86	11
4.	46	2	132	10	69	19	71	9
5.	23	1	119	9	95	26	118	15
6.	69	3	145	11	65	18	157	20
7.	92	4	66	5	109	30	118	15
8.	185	8	106	8	95	26	78	10
9.	277	12	79	6	76	21	78	10
10.	92	4	132	10	87	24	157	20
11.	46	2	92	7	142	39	71	9
12.	92	4	132	10	98	27	94	12

[a]The index is 100 times the monthly marriages or conceptions divided by the annual average monthly marriages or conceptions.

Source: Friends Historical Library, Swarthmore College, Swarthmore, Pennsylvania.

TABLE 4. Chester County Wills, Transfer of Property, 1750s and 1790s

		Nuclear Kin[a]		*Near Kin*[b]		*Distant Kin*[c]		*Non-kin*[d]	
1750s	Total *N*	N	%	N	%	N	%	N	%
Females	33	20	61	22	67	8	24	5	15
Males	25	24	96	11	44	2	8	4	16
1790s									
Females	50	35	70	38	76	13	26	8	16
Males	33	28	85	11	33	4	12	5	15
% Change									
Females		+9		+9		+2		+1	
Males		−8		−11		+4		−1	

[a]Includes sons and daughters
[b]Includes sisters, mothers, fathers, granddaughters, grandsons, sons-in-law, daughters-in-law, brothers-in-law, sisters-in-law
[c]Includes nieces, nephews, cousins
[d]Includes friends, charity

Source: Chester County Courthouse, West Chester, Pennsylvania.

TABLE 5. Chester County Wills, Executors, 1750s and 1790s

Sole Executors 1750s	*Total* N	*Wife* N	%	*Daughter* N	%	*Son* N	%	*Son-in-Law* N	%	*Male Friend* N	%
Female	34	—	—	0	0	5	15	1	3	6	18
Male	23	2	9	0	0	2	19	0	0	3	13
1790s											
Female	50	—	—	3	6	10	20	7	14	20	40
Male	33	5	15	1	3	11	33	0	0	13	39
% Change											
Female				+6		+5		+11		+22	
Male		+6		+3		+24				+26	

Source: Chester County Courthouse, West Chester, Pennsylvania.

TABLE 6. New Castle County Households, 1782

Average Number of White Persons Per House

	Number of Households	Population	Average Per Household
Brandywine	258	1,366	5.29
Christiana	692	3,193	4.61
Average	950	4,559	4.80

Female-Headed Households

	Number of Households	Female-Headed	Percent of Households
Brandywine	258	6	2
Christiana	692	77	11

Composition of Female-Headed Households, Christiana

	Total N	%	N Age Group in Population
Females 18 and under	78	9	827
Females over 18	119	15	771
Males 18 and under	45	5	891
Males over 18	14	2	704
Total	256	8	3193

Source: Census of 1782 for New Castle County, Delaware Hall of Records, Dover, Delaware.

TABLE 7. Chester County Households, 1783

Average Number of Persons Per Household[a]

	Number of Houses	Population	Average Per House
Birmingham	84	543	6.46
East Bradford	89	568	6.38
Kennett	99	617	6.23
New Garden	70	541	7.73
Pennsbury	76	534	7.02
West Bradford	101	591	5.85

Female-Headed Households

	Number of Houses	Female-Headed	Percent Female-Headed	Total Acres Owned
Birmingham	84	7	12	250
East Bradford	89	2	2	250
Kennett	99	0	0	0
New Garden	70	4	6	375
Pennsbury	76	0	0	0
West Bradford	101	5	5	215

Size of Household[b]

	Total N	1–4	%	5–9	%	10–14	%	15–20	%
Birmingham	89	23	26	56	63	10	11	0	0
East Bradford	86	25	29	49	57	12	14	0	0
Kennett	68	14	21	26	38	19	28	9	13
New Garden	82	22	27	44	54	16	20	0	0

[a]Twenty-one blacks lived in these six townships, fourteen in Kennett.
[b]Includes one to four houses.

Source: Chester County Courthouse, West Chester, Pennsylvania.

TABLE 8. New Castle County Households, 1850

	Brandywine (N = 64)	Christiana (N = 83)	Total (N = 147)
Persons in household	5.4	5.8	5.6
Related persons in household	4.7	4.8	4.8
Persons in nuclear families	4.5	4.5	4.5
Female-headed	14%	11%	12%
Black	8%	6%	7%
Farmers	34%	28%	31%
Laborers	14%	19%	16%
Literate	94%	87%	90%
Male children 15+	78%	75%	76%
Female children 15+	80%	80%	80%
Non-kin female 10+	81%	77%	79%
Farm Households	Brandywine (N = 22)	Christiana (N = 23)	Total (N = 45)
Persons in household	5.8	6.3	6.0
Related persons in household	5.2	4.3	4.8
Persons in nuclear families	5.1	4.0	4.6
Female-headed	0	13%	6.5%
Black	0	0	0
Literate	100%	100%	100%
Male children 15+	77%	78%	78%
Female children 15+	64%	74%	69%
Non-kin female 10+	86%	52%	69%
Nonfarm Households	Brandywine (N = 42)	Christiana (N = 60)	Total (N = 102)
Persons in household	5.2	5.7	5.5
Related persons in household	4.4	5.0	4.5
Persons in nuclear families	4.2	4.8	4.5
Female-headed	21%	10%	15.5%
Black	12%	8%	10%
Laborers	21%	27%	24%
Literate	90%	82%	86%
Male children 15+	77%	73%	75%
Female children 15+	88%	82%	84%
Non-kin female 10+	86%	82%	83%

Source: United States, Population Census, 1850. Computer data from 10 percent sample supplied by Jack Michel, American Studies Program, University of Delaware.

TABLE 9. Chester and New Castle Counties Household Inventories, 1750s

	Chester Female-Headed (N = 40)		Chester Male-Headed (N = 45)		New Castle Male-Headed (N = 42)		Total N (N = 87)	Percent
	N	%	N	%	N	%		
Animals								
Cows	20	50	34	76	30	71	84	66
Sheep	6	15	25	56	21	50	52	41
Hogs	4	10	24	53	23	55	51	40
Field Tools								
Sickles	2	15	16	36	17	40	35	28
Scythes	1	3	17	38	19	45	37	29
Cradles	1	3	4	9	7	17	12	9
Scythes and hangings	0	0	2	4	0	0	2	2
Women Workers								
Enslaved	0	0	4	9	8	19	12	9
Indentured	2	5	2	4	5	12	9	7
Textile Materials								
Wool	3	8	6	13	9	21	18	14
Flax	5	13	10	22	19	45	34	27
Flax seed	1	3	9	20	8	19	18	14
Textile Tools								
Hatchels	6	15	17	38	11	26	34	27
Cards	6	15	6	13	9	21	21	17
Linen wheels	18	45	29	64	17	40	64	50
Wool wheels	4	10	15	33	23	55	42	33
Looms	0	0	5	11	9	21	14	11
Irons	5	13	5	11	4	10	14	11
Processed Textiles								
Yarn	8	20	13	29	17	40	38	30
Yardage	18	45	21	47	20	48	59	46
Finished Textiles								
Beds	32	80	43	96	31	74	106	83
Featherbeds	19	48	8	18	18	43	45	35
Chaff and flock	6	15	10	22	9	21	25	20
Pillows	10	25	3	7	7	17	20	16
Bolsters	15	38	5	11	4	10	24	19
Sheets	21	53	10	22	14	33	45	35
Pillowcases	11	28	8	18	8	19	27	21
Blankets	12	30	8	18	11	26	31	24
Coverlids	12	30	4	9	7	17	23	18
Quilts	3	8	1	2	5	12	9	7
Curtains	9	23	4	9	3	7	16	13
Rugs	8	20	5	11	0	0	13	10
Table linen	17	43	15	33	10	24	42	33
Food-Processing Tools								
Tongs and racks	22	55	37	82	31	74	90	71
Pots	22	55	34	76	26	62	82	65
Pails	10	25	12	27	12	29	34	27
Tubs	9	23	17	38	17	40	43	34
Frying pans	14	35	22	49	20	48	56	44
Kettles	17	43	12	27	13	31	42	33
Griddles and gridirons	5	13	13	29	20	48	38	30

	Chester Female-Headed (N = 40)		Chester Male-Headed (N = 45)		New Castle Male-Headed (N = 42)		Total N (N = 87)	Percent
	N	%	N	%	N	%		
Baskets	5	13	2	4	0	0	7	6
Dough troughs	5	13	27	60	9	21	41	32
Churns	8	20	14	31	12	29	34	27
Cheese presses	2	5	8	18	5	12	15	12
Woodenware	9	23	11	24	15	36	35	28
Earthenware	18	45	23	51	25	60	66	52
Tinware	13	33	8	18	9	21	26	20
Plate	24	60	33	73	26	62	83	65
Stoves and ovens	2	5	0	0	0	0	2	2
Storage								
Chests	23	58	38	84	29	69	90	71
Cases	14	35	13	29	8	19	35	28
Trunks	11	28	8	18	6	14	25	20
Cupboards	6	15	7	16	6	14	19	15
Clothespresses	3	8	7	16	2	5	12	9
Service								
Chairs	23	58	36	80	27	64	86	68
Tables	19	48	36	80	28	67	83	65
nives and forks	9	23	12	27	19	45	40	32
Spoons	15	38	10	22	13	31	38	30
Special china	11	28	7	16	14	33	32	25
Candlesticks	14	35	14	31	13	31	41	32
Looking glasses	19	48	17	38	17	40	53	42
Desks	2	5	9	20	6	14	17	13
Bedsteads	11	28	3	7	15	36	29	23
Warming pans	17	43	14	31	6	14	37	29
Clocks and watches	7	18	12	27	1	2	20	16
Pictures	1	3	0	0	1	2	2	2
Books	15	38	28	62	24	57	67	53

Source: Chester County Courthouse, West Chester, Pennsylvania, and Delaware Hall of Records, Dover, Delaware.

TABLE 10. Chester and New Castle Counties Household Inventories, 1790s

	Chester				New Castle				Total	
	Male-Headed		Female-Headed		Male-Headed		Female-Headed		Male-Headed	
	(N = 43)		(N = 43)		(N = 34)		(N = 33)		(N = 77)	
	N	%	N	%	N	%	N	%	N	%
Animals										
Cows	36	84	17	40	28	82	19	58	64	83
Sheep	29	67	12	28	17	50	7	21	46	60
Hogs	30	70	8	19	15	44	5	15	45	58
Field Tools										
Sickles	11	26	0	0	7	21	1	3	18	23
Scythes	16	37	0	0	13	38	1	3	29	38
Cradles	8	19	0	0	6	18	0	0	14	18
Scythes and hangings	10	23	1	2	1	3	1	3	11	14
Female Workers										
Enslaved	1	2	0	0	14	41	4	12	15	19
Indentured	0	0	1	2	2	6	0	0	2	3
Textile Materials										
Wool	6	14	6	14	4	12	3	9	10	13
Flax	13	30	7	16	17	50	8	24	30	39
Flax seed	6	14	1	2	9	26	1	3	15	19
Textile Tools										
Hatchels	15	35	9	21	12	35	7	21	27	35
Cards	5	12	6	14	8	24	4	12	13	17
Linen wheels	28	65	26	60	24	71	18	55	52	68
Wool wheels	24	56	12	28	14	41	11	33	38	49
Looms	6	14	0	0	4	12	0	0	10	13
Irons	14	33	14	33	16	47	16	48	30	39
Processed Textiles										
Yarn	8	19	16	37	9	26	7	21	17	22
Yardage	14	33	18	42	12	35	9	27	26	34
Finished Textiles										
Beds	43	100	38	88	28	82	29	88	71	92
Featherbeds	19	44	18	42	11	32	19	58	30	39
Chaff and other	18	42	4	9	5	15	4	12	23	30
Pillows	3	7	10	23	3	9	6	18	6	8
Bolsters	4	9	11	26	5	15	10	30	9	12
Sheets	12	28	22	51	17	50	20	61	29	38
Pillowcases	7	16	10	23	9	26	14	42	16	21
Blankets	10	23	15	35	14	41	15	45	24	31
Coverlids	13	30	17	40	11	32	11	33	24	31
Quilts	1	2	10	23	4	12	11	33	15	6
Curtains	4	9	8	19	10	29	14	42	14	18
Rugs	3	7	3	7	5	15	5	15	8	10
Table linen	10	23	20	47	12	35	18	55	22	29
Food-Processing Tools										
Tongs and racks	32	74	24	56	25	74	24	73	57	74
Pots	34	79	23	53	22	65	23	70	56	73
Pails	16	37	14	33	10	29	12	36	26	34
Tubs	20	47	15	35	13	38	15	45	33	43
Frying pans	26	60	15	35	20	59	13	39	46	60
Kettles	28	65	18	42	21	62	20	61	49	64

	Chester				New Castle				Total	
	Male-Headed		Female-Headed		Male-Headed		Female-Headed		Male-Headed	
	(N = 43)		(N = 43)		(N = 34)		(N = 33)		(N = 77)	
	N	%	N	%	N	%	N	%	N	%
Griddles and gridirons	21	49	7	16	20	59	17	52	41	53
Baskets	10	23	7	16	5	15	8	24	15	19
Dough troughs	24	56	9	21	11	32	6	18	35	45
Churns	24	56	6	14	7	21	8	24	31	40
Cheese presses	3	7	2	5	2	6	1	3	5	6
Woodenware	11	26	7	16	19	56	10	30	30	39
Earthenware	22	51	18	42	18	53	22	67	40	52
Tinware	18	42	11	26	12	35	10	30	30	39
Plate	28	65	35	81	22	65	27	82	50	65
Stoves and ovens	21	49	14	33	15	44	13	39	36	47
Storage										
Chests	30	70	27	63	27	79	20	61	57	74
Cases	22	51	18	42	15	44	12	36	37	48
Trunks	5	12	4	9	13	38	9	27	18	23
Cupboards	18	42	7	16	20	59	13	39	38	49
Clothespresses	5	12	5	12	0	0	4	12	5	6
Service										
Chairs	39	91	33	77	30	88	32	97	69	90
Tables	41	95	33	77	28	82	31	94	69	90
Knives and forks	11	26	12	28	18	53	15	45	29	38
Spoons	12	28	11	26	14	41	19	58	26	34
Candlesticks	12	28	11	26	15	44	13	39	27	35
Looking glasses	21	49	19	44	21	62	24	73	42	55
Desks	16	37	6	14	18	53	9	27	34	44
Bedsteads	25	58	22	51	19	56	22	67	44	57
Warming pans	10	23	12	28	7	21	11	33	17	22
Clocks and watches	19	44	5	12	8	24	11	33	27	35
Pictures	0	0	0	0	4	12	6	18	4	5
Books	29	67	21	49	24	71	21	64	53	69

Source: Chester County Courthouse, West Chester, Pennsylvania, and Delaware Hall of Records, Dover, Delaware.

TABLE 11. Chester and New Castle Counties, Value of Estates and Wearing Apparel, Heads of Household, 1750s and 1790s

Estate in Pounds	Chester Male-Headed 1750	Chester Male-Headed 1790	Chester Female-Headed 1750	Chester Female-Headed 1790	New Castle Male-Headed 1750	New Castle Male-Headed 1790	New Castle Female-Headed 1790
0–9	1	0	0	1	1	0	0
10–29	1	4	5	3	3	2	2
30–49	7	7	8	4	3	2	5
50–99	8	1	15	6	10	3	10
100–199	15	12	7	9	10	5	14
200–499	11	9	6	6	8	12	9
500+	3	14	2	4	3	11	3
Total	46	47	43	33	38	35	43
Apparel in Pounds							
0–4	9	10	3	5	17	7	4
5–9	10	8	13	6	9	12	11
10–19	11	6	11	4	7	4	7
20–29	5	3	3	6	1	2	3
30+	1	6	1	1	1	1	2
Total	36	33	31	22	35	26	27

Source: Chester County Courthouse, West Chester, Pennsylvania, and Delaware Hall of Records, Dover, Delaware.

TABLE 12. Chester and New Castle Counties, Poorhouse Women

Chester County by Ethnic Group, 1800–1850

		Female		European Female		African Female	
Date	Total N	N	%	N	%	N	%
1800	84	44	52	43	98	1	2
1801	57	33	58	29	88	4	12
1806	74	35	47	35	100	0	0
1807	89	37	42	35	100	0	0
1809	131	67	51	67	100	0	0
1812	116	50	43	50	100	0	0
1813	114	48	42	48	100	0	0
1814	119	64	54	64	100	0	0
1815	123	57	46	57	100	0	0
1821	292	40	48	140	100	0	0
1822	272	131	48	113	84	18	16
1823	287	159	55	159	100	0	0
1843	272	136	50	124	77	12	23
1844	284	144	51	130	70	14	30
1846	290	144	50	136	85	8	15
1847	302	132	44	119	74	13	24
1850	337	174	52	168	85	6	15
Total	3243	1595	49	1517	95	76	5

New Castle County, by Ethnic Group and Cause of Admission, 1833–1835

Ethnic Group	Number	Mental[a]	Age	Poverty	Pregnant	Physical	Unknown
European	54	17	4	5	7	19	2
Irish	20	3	1	1	1	13	1
African	13	3	1	1	1	7	0
Total	87	23 (26%)	6 (7%)	7 (8%)	9 (10%)	39 (45%)	3 (3%)

[a]Includes terms *insanity*, *simple*, *idiocy*

Source: Poorhouse Records, Chester County Courthouse, West Chester, Pennsylvania, and Delaware Hall of Records, Dover, Delaware.

TABLE 13. Relative Wholesale Butter Prices, Philadelphia, 1784–1849

Year	Price	Year	Price	Year	Price
1784	133.79	1806	167.79	1828	81.16
1785	120.56	1807	139.20	1829	83.48
1786	107.94	1808	115.22	1830	76.96
1787	100.93	1809	125.23	1831	102.19
1788	82.71	1810	145.92	1832	114.60
1789	70.56	1811	145.93	1833	116.07
1790	99.53	1812	126.37	1834	95.04
1791	107.71	1813	127.21	1835	124.38
1792	106.82	1814	180.41	1836	171.16
1793	128.68	1815	212.79	1837	135.62
1794	174.06	1816	211.03	1838	146.13
1795	165.65	1817	170.32	1839	151.39
1796	142.98	1818	146.56	1840	100.93
1797	123.22	1819	131.00	1841	88.84
1798	161.49	1820	102.82	1842	78.22
1799	160.86	1821	100.51	1843	66.26
1800	148.03	1822	131.00	1844	78.46
1801	146.77	1823	95.05	1845	99.99
1802	118.59	1824	81.79	1846	96.93
1803	95.04	1825	91.68	1847	116.07
1804	174.27	1826	105.13	1848	118.49
1805	179.15	1827	94.83	1849	97.14

Base: $.09908 per pound in firkins or kegs.
Source: Anne Bezanson, Robert D. Gray, and Miriam Hussey, *Wholesale Prices in Philadelphia, 1784–1861* (Philadelphia: Univesity of Pennsylvania Press, 1937), 2:25.

TABLE 14. Agricultural Products, Pennsylvania, 1840

Amount by Bushels	
Oats	20,641,819
Corn	14,240,022
Wheat	13,213,077
Potatoes	9,535,663
Rye	6,613,873
Value	
Dairy products	$3,187,292
Orchard	618,179
Market	232,912

Source: Samuel Bulkley Ruggles, *Tabular Statements from 1840 to 1870, of the Agricultural Products of the States and Territories of the United Staes of America* (New York: Chamber of Commerce, 1874).

TABLE 15. Butter-producing Farms in the Philadelphia Hinterland by Annual Production Category, 1850

	Annual Production Category					
County	8,000 lbs. and over	2,000 to 7,999 lbs.	600 to 1,999 lbs.	200 to 599 lbs.	under 200 lbs.	Total
Delaware	2	232	452	429	292	1,407
Percent	0.1	17	32	30	21	100
Chester	1	131	803	2,589	1,237	4,761
Percent	0.02	3	17	54	26	100
New Castle	3	46	358	812	470	1,689
Percent	0.2	3	21	48	28	100
Total	6	409	1,613	3,830	1,999	7,859
Percent	0.08	5	21	48	25	100

Source: United States, Census of Agriculture, 1850.

TABLE 16. New Castle County, Brandywine Hundred, Agricultural Products, 1850

Value	
Farm	$5,028.13
Per acre	63.19
Stock	385.70
Machinery	234.38
Slaughtered goods	63.34
Market produce	24.66
Orchard produce	16.17
Production	
Butter (lbs.)	630.39
Corn (bu.)	172.58
Wheat (bu.)	104.06
Oats (bu.)	82.94
Irish potatoes (bu.)	67.89
Buckwheat (bu.)	5.18
Hay (tons)	19.53
Number of milk cows	6.47
Percent of farms with milk cows	74.42

N = 64

Source: United States, Census of Agriculture, 1850, University of Delaware, American Studies Program Data Set.

TABLE 17. Chester County, Farm Outbuildings, 1978

Type	Township	N Farms	N Type	Percent with Outbuildings	Average Size (sq. ft.)
Spring	Concord	79	17	25	186
Spring	Bradford	228	57	25	160
Spring	Edgemont	44	23	52	127
Milk	West Caln	115	51	54	84
Barns	West Caln	115	71	62	
Barns	Birmingham	43	39	91	
Barns	Sadsbury	134	55	41	

Source: 1798 Tax Lists, Chester County Historical Society, West Chester, Pennsylvania.

TABLE 18. Quaker Women Ministers, 1700–1799

	Died 1700–1749 (N=69)		Died 1750–1799 (N=143)	
	Mean	N	Mean	N
Age at marriage	32	10	26	49
Age at entering ministry	35	14	33	57
Number of years in ministry	24	25	31	45
Age at death	62	25	67	109
	%	N	%	N
Born in Great Britain	82	34	40	68
Born in America	15	34	60	68

Source: Willard Heiss, ed., *Quaker Biographical Sketches of Ministers and Elders, and Other Concerned Members of the Yearly Meeting of Philadelphia* (Indianapolis, Ind.: n.p., 1972); John Smith, "Lives of Ministers among Friends," vol. 1, Quaker Collection, Haverford College; J. W. Lippincott, "Society of Friends, Deceased Ministers and Elders, List Compiled from Friends' Memoirs and Journals, 1656–1898," Quaker Collection, Haverford College; Journals and Diaries at Friends Historical Library, Swarthmore College, and Quaker Collection, Harverford College.

TABLE 19. Chester County, White Females and Males Signing Wills, 1729–1850

	Women					Men				
	Sign.		Mark		Total	Sign.		Mark		Total
	N	%	N	%	N	N	%	N	%	N
1729–44	16	46	19	54	35	188	74	65	26	253
1745–54	8	19	35	81	43	207	73	78	27	285
1750–59	9	26	26	74	35	18	82	4	18	22
1755–64	12	27	33	73	45	200	69	88	31	288
1765–74	33	44	42	56	75	248	73	90	27	338
1790–99	25	50	23	46	48	22	67	11	33	33
1840	(Total N=57,515; illiterate female + male=751 or 1%)									
1850	(Total N=61,215; illiterate female=1,318 or 2% of total, 4% of female population)									

Source: Chester County Courthouse, West Chester, Pennsylvania.

TABLE 20. Chester and New Castle Counties, Literacy among Black Females Twenty and over, 1850

	Total N	Literate N	Literate %
Chester County			
Penn	12	9	75
Kennett	50	26	52
New Garden	61	27	44
West Chester	121	82	68
Total	244	144	59
New Castle County			
Brandywine	107	42	39
Christiana	65	29	45
White Clay	90	82	91
Total	252	149	59

By State of Birth	Pennsylvania			Md., Del., Unknown		
	Total	Literate		Total	Literate	
	N	N	%	N	N	%
Chester County						
Penn	6	6	100	6	3	50
Kennett	23	16	70	26	9	35
New Garden	21	15	71	36	13	36
West Chester	85	55	65	23	17	74
Total	135	92	68	90	42	47

Source: United States Manuscript Census, 1850.

TABLE 21. Chester and New Castle Counties, Female Children, 5–15, in School Previous Year, 1850

	Black			White		
	Total	In School		Total	In School	
	N	N	%	N	N	%
Chester County						
Penn	11	6	55	91	58	64
Kennett	29	11	38	92	77	84
New Garden	44	26	59	139	117	84
West Chester	65	32	49	312	237	76
Total	149	75	50	634	489	77
New Castle County						
Brandywine	55	0	0	486	390	80
Christiana	46	4	9	622	390	63
White Clay	51	6	12	241	180	75
Total	152	10	7	1349	960	71

Source: United States Manuscript Census, 1850.

TABLE 22. New Castle County, Educational Provisions in Indentures, 1818–1839

	Black (N=113)		White (N=264)	
	N	%	N	%
No schooling	43	38	41	16
Reading	22	19	10	4
Reading and writing	36	32	121	46
Reading, writing, and arithmetic	0	0	18	7
School time specified	13	12	74	28

N=378
Source: Indenture Records, Hall of Records, Dover, Delaware.

TABLE 23. Black Population of Chester and New Castle Counties as a Percent of Total Population, 1850

County	Total Population	Number		Number		Percent Black
		Black	Female	White	Female	
Chester	66,438	5,223	2,572	61,215	30,452	8
New Castle	42,780	8,015[a]	4,000	34,765	17,517	19
Total	109,218	13,238	6,572	95,980	47,969	12
Selected Chester Townships						
West Goshen	940	248	112	692	347	26
New Garden	1,391	304	148	1,087	553	22
Pocopsin	592	95	40	497	255	16
Upper Oxford	1,021	158	79	863	405	15
West Marlboro	1,130	166	86	964	488	15
Westtown	789	120	68	669	349	15
West Chester	3,172	451	240	2,721	1,474	14
Kennett	1,706	222	105	1,484	779	13
London Grove	1,425	172	78	1,253	621	12
West Bradford	1,585	183	86	1,402	673	12
Charlestown	979	112	50	867	435	11
New London	2,042	216	102	1,826	904	11
East Whiteland	1,194	127	65	1,067	543	11
Total	17,966	2,574	1,259	15,392	7,826	14
Black women as percent of blacks in county: 49						
Black women as percent of females in county: 14						

[a]Includes 394 enslaved, 187 women.
Source: United States, Population Census, 1850

TABLE 24. Prohibition Petitions by County, 1848

County	Total Number	Percent Male
Washington	20	85
Allegheny	15	73
Delaware	14	100
Chester	13	62
Butler	6	100
Carbon	3	100
Luzerne	3	100
Adams	3	100
Lancaster	3	33
Columbia	2	0
Bucks	2	0
Mifflin	1	100
Schuylkill	1	100
Cambria	1	0
Union	1	100
Northumberland	1	100
Blair	1	100
Centre	1	0

N = 93
Source: Pennsylvania Archives, Harrisburg, Pennsylvania.

TABLE 25. Characteristics of Women's Rights Households, 1850

Ethnicity	Number	White	Black	Irish
Women's rights	24	24	0	0
Kennett farms	127[a]	125	2	0
Kennett Township	104[b]	87	13	0
West Chester	100[c]	77	15	8
Female-Headed				
Women's rights	20	5	0	0
Kennett farms	127	9	0	0
Kennett Township	104	11	0	0
West Chester	100	17	2	1

Women Working Out (14 to 45)	Number	Percent
Women's rights	NA	NA
Kennett farms	46	24
Kennett Township	24	17
West Chester	65	38
Mean Income	Number	Amount
Women's rights	18	$10,306
Kennett farms	94	6,753
Kennett Township	44	5,186
West Chester	54	5,953

Households with Live-in Female Servants (August)	Total Number	Number with Servants	Percent
Women's rights	20	14	70
Kennett farms	127	43	34
Kennett Township	104	20	19
West Chester	100	43	43
Mean Age by Gender and Age of Children at Home[d]	Total Number	Male	Female
Women's rights	20	12.25	15.74
Kennett farms	127	12.54	12.60
Kennett Township	104	9.88	10.60
West Chester	100	9.00	9.39

[a]All farms in township
[b]Every third household of 312 households
[c]Every fifth household of 504 households
[d]All under 40 with same surname as heads of household plus others under 14
Source: United States, Manuscript Census, 1850.

NOTE ON SOURCES

When I began this study, scholars warned there might not be sufficient sources to gain any clear idea of the lives of rural women. Instead, I found the sources so extensive that I could use only a part of them. To cover one hundred years of women's rural life, even in one small geographical area, did present a challenge. For the period from 1750 to 1800, I relied heavily on quantitative sources: church records, wills, inventories, tax lists, and censuses. For the period from 1800 to 1850, many literary sources were available, allowing me to draw on diaries, letters, and journals, as well as early imprints and almanacs.

Four collections contained the bulk of both quantitative and literary sources: those of the Chester County Courthouse, the Chester County Historical Society, the Friends Historical Library at Swarthmore College, and the Quaker Collection at Haverford College. The Delaware records are rich but difficult to use. Microfilmed and reproduced copies of inventories and wills, together with poorhouse and indenture records, enabled me to document New Castle County more extensively than I had hoped. There are many more records in the Delaware Hall of Records to be utilized. Samples of inventories and wills from New Castle and Chester counties indicate they are fruitful sources for the historian willing to spend time becoming familiar with them. Legal documents also were so extensive at the Chester County Courthouse that I could sample only a portion. The Esther Lewis collection, now housed in the Friends Historical Library at Swarthmore College, is the most complete record of a rural woman in Chester County yet available.

Other collections at the Eleutherian Mills–Hagley Library, Historical

Society of Pennsylvania, Historical Society of Delaware, Painter Library, the New-York Historical Society, and the National Archives contain scattered treasures that allowed me to develop several approaches to the lives of rural women. Many letters and journals from the early nineteenth century still remain in private hands, but even those available have been underutilized by historians.

Historians, myself included, still find material objects awkward to use effectively. Curators at the Henry Ford Museum, the Mercer Museum, and the Smithsonian, as well as countless smaller museums, helped me understand the tools and skills of rural women. During the course of this effort to understand material culture, I became convinced that the artifacts of the age of wood, earthenware, and homespun need far more systematic analysis by historians than they have received.

Early imprints were an important source for this study, and the Hagley Library and the National Agriculture Library provided the best collections. Almanac collections at the Friends Historical Library and the Hagley Library provided me with new ways to look at the literature of the time. Newspapers were used less extensively primarily because there were so many other sources available rather than because of a lack of reporting on the activities of women.

Secondary sources are cited in the notes. I relied heavily on a wide variety of dissertations and recently published articles that contributed important insight to the still largely unwritten agricultural and rural history of the Mid-Atlantic.

Abbreviations

AH	*Agricultural History*
AQ	*American Quarterly*
AHR	*American Historical Review*
CCCH	Chester County Courthouse, West Chester, Pennsylvania
CCHS	Chester County Historical Society, West Chester, Pennsylvania
DH	*Delaware History*
DHR	Delaware Hall of Records, Dover, Delaware
EHR	*Economic History Review*
EMHL	Eleutherian Mills, Hagley Library, Greenville, Delaware
FS	*Feminist Studies*
FHLSC	Friends Historical Library, Swarthmore College, Swarthmore, Pennsylvania
FLMM	Fonthill Library, Mercer Museum, Doylestown, Pennsylvania
FL	*Friends Library*
HSD	Historical Society of Delaware, Wilmington, Delaware
HSP	Historical Society of Pennsylvania, Philadelphia
JAF	*Journal of American Folklore*
JEH	*Journal of Economic History*
JHF	*Journal of the History of the Family*

JIH	*Journal of Interdisciplinary History*
JMF	*Journal of Marriage and the Family*
JNH	*Journal of Negro History*
JSH	*Journal of Social History*
NA	National Archives, Washington, D.C.
NYHS	New-York Historical Society, New York City
PA	*Pennsylvania Archaeologist*
PA	Pennsylvania Archives, Harrisburg, Pennsylvania
PLTA	Painter Library, Tyler Arboretum, Lima, Pennsylvania
PH	*Pennsylvania History*
PMHB	*Pennsylvania Magazine of History and Biography*
PS	*Population Studies*
QCHC	Quaker Collection, Haverford College, Haverford, Pennsylvania
QH	*Quaker History*
WML	Winterthur Museum Library, Winterthur, Delaware
WP	*Winterthur Portfolio*
WMQ	*William and Mary Quarterly*

NOTES

CHAPTER 1

1. Henry Francis James, *The Agricultural Industry of Southeastern Pennsylvania: A Study in Economic Geography* (Philadelphia: n.p., 1928), pp. 2–14; Henry Drinker Biddle, *Extracts from the Journal of Elizabeth Drinker from 1759 to 1807* A.D. (Philadelphia: Lippincott, 1889), p. 374.

2. Jack Michel, "In a Manner and Fashion Suitable to Their Degree: A Preliminary Investigation of the Material Culture of Early Pennsylvania," *Working Papers from the Regional Economic History Center* 5, no. 1 (1981):1–83.

3. Peter Kalm, *Travels in North America*, ed. Adolph B. Benson, 2 vols. (New York: Dover; 1966, reprint of 1937 ed.), 2:267–274.

4. Willard Heiss, ed., *Quaker Biographical Sketches of Ministers and Elders, and Other Concerned Members of the Yearly Meeting of Philadelphia* (Indianapolis, Ind., 1972), pp. 122, 159; John F. Watson, "Memorials of Country Towns and Places in Pennsylvania," Historical Society of Pennsylvania, *Memoirs* 2 (1830):171.

5. Wilmer W. MacElree, *Along the Western Brandywine*, 2d ed. (West Chester, 1912), pp. 107–108; purchases listed in Diary of Richard Barnard, HSP, 1–15–1780, 9–10–1782, and 12–4–1788.

6. Marshall J. Becker, "Lenape Archaeology: Archaeological and Ethnohistoric Considerations in Light of Recent Excavation," *PA* 50 (1980):21; and Anthony F. C. Wallace, "Woman, Land and Society: Three Aspects of Aboriginal Delaware Life," *PA* 17 (1947):1–35. For the debate over lineage, see C. A. Weslager, *The Delaware Indian Westward Migration* (Wallingford, Pa.: Middle Atlantic Press, 1978), p. 91, who cites an 1821 document in which they are described as matrilineal and A. R. Dunlap and C. A. Weslager, "Contributions to the Ethno-History of the Delaware Indians on the Brandywine," *PA* 30 (1960):18–21.

7. Susan Klepp, "Five Early Pennsylvania Censuses," *PMHB* 106 (1982):483–514; "Reincke's Journal of a Visit among the Swedes of West Jersey, 1745," *PMHB* 33 (1909):10.

8. Barry Levy, "The Light in the Valley: The Chester and Welsh Tract Quaker Communities and the Delaware Valley, 1681–1750" (Ph.D. diss., University of Pennsylvania, 1976), pp. 51–63; Charles H. Browning, *Welsh Settlement of Pennsylvania* (Philadelphia: Campbell, 1912), pp. 421–425, 491.

9. James Thomas Lemon, "The Agricultural Practices of National Groups in Eighteenth Century Southeastern Pennsylvania," *Geographical Review* 56 (1966):469; CCCH, Wills, #1651 Mary Davis, 1753.

10. Gerelyn Hollingsworth, "Irish Quakers in Colonial Pennsylvania: A Forgotten Segment of Society," *Journal of Lancaster County Historical Society* 79 (1975):150–162.

11. Albert Cook Myers, *Immigration of the Irish Quakers into Pennsylvania, 1682–1750* (Swarthmore, Pa.: n.p., 1902), has genealogies for Ann Calvert Hollingsworth and Elizabeth Dick Harlan.

12. Watson, "Memorials," 172; CCCH, Wills, #765 Mary Harlan, #1443 Enoch Hollingsworth; Alpheus H. Harlan, comp., *History and Genealogy of the Harlan Family* (n.p., 1914).

13. Ibid.

14. Myers, *Immigration*, p. 187; CCCH, Wills, #464 Aaron Harlan, #1089 Sarah Harlan. According to Lemon, "Agricultural Practices," p. 469, Scotch-Irish and Scots-Irish surnames accounted for almost one quarter of the Chester County population in 1759. The Scots, however, were on the western frontier or occupied scattered areas among the dominant English. He gives the following ethnic proportions: English, 59 percent; Scotch-Irish, Scots-Irish, 23 percent; Welsh, 8 percent; German-speaking, 5 percent; other, 5 percent; total 24,500.

15. Charles Knowles Bolton, *Scotch Irish Pioneers in Ulster and America* (Boston: Bacon & Brown, 1910), pp. 267–282. Native Irish immigration into Pennsylvania is still poorly documented. David Noel Doyle, *Ireland, Irishmen and Revolutionary America, 1760–1820* (Dublin: Mercier, 1981), and Edward C. Carter II, "A 'Wild Irishman' under Every Federalist's Bed: Naturalization in Philadelphia, 1789–1806," *PMHB* 94 (1970):331–346, have established the importance of their numbers and political activities but say little of their culture and even less about the women.

16. Irene G. Shur, "Emergence of the Free Common Schools in Chester County, Pennsylvania: 1834–1874" (Ph.D. diss., University of Pennsylvania, 1976), pp. 11, 17, 29, 259. Reformed churches were founded in East Coventry in 1743 and East Vincent in 1758, a Lutheran church in Pikeland, and Mennonite churches in East Coventry and Phoenixville. A similar dispersed settlement occurred across the border in Maryland. See Elizabeth Kessel, " 'A Mighty Fortress Is Our God': German Religious and Educational Organizations on the Maryland Frontier, 1734–1800," *Maryland Historical Magazine* 77 (Winter 1982):378–387, and Elizabeth Augusta Kessel, "Germans on the Maryland Frontier: A Social History of Frederick County, Maryland, 1730–1800" (Ph.D. diss., Rice University, 1981).

17. Anne Catherine Bieri Hebert, "The Pennsylvania French in the 1790's: The Story of Their Survival" (Ph.D. diss., University of Texas, Austin, 1981), pp. 3–29; J. P. Brissot de Warville, *New Travels in the United States of America, 1788*, ed. Durand Echeverria (Cambridge, Mass.: Harvard University Press, 1964), pp. 203–204.

18. Charles L. Coleman, "The Emergence of Black Religion in Pennsylvania, 1776–1850," *Pennsylvania Heritage* 4 (1977):24–28; Lewis V. Baldwin, " 'Invisible' Strands in African Methodism: A History of the African Union Methodist Protestant and Union American Methodist Episcopal Churches, 1805–1980" (Ph.D. diss., Northwestern University, 1980); Jean R. Soderlund, "Black Women in Colonial Pennsylvania," *PMHB* 107 (1983):57–58. There was a large minority of mulattos, 13 percent, within the black population because of a rural gender imbalance earlier in the century. Carl D. Oblinger, "Freedom Foundations: Black Communities in Southeastern Pennsylvania Towns, 1780–1860," *Northwest Missouri State University Studies* 33 (November 1972):3–23 and "Alms for Oblivion: The Making of a Black Underclass in Southeastern Pennsylvania, 1780–1860," in *The Ethnic Experience in Pennsylvania*, ed. John E. Bodnar (Lewisburg, Pa.: Bucknell University Press, 1973), pp. 94–119; Carol V. R. George, *Segregated Sabbaths: Richard Allen and the Emergence of Independent Black Churches, 1760–1840* (New York: Oxford University Press, 1973), pp. 54–62.

19. W. Emerson Wilson, *Mount Harmon Diaries of Sidney George Fisher, 1837–1850* (Wilmington: Historical Society of Delaware, 1976), p. 70.

20. These estimates are based on families reconstituted from the fragmentary church records, which no doubt underrepresented children and older people. Jack D. Marietta, "Ecclesiastical Discipline in the Society of Friends, 1682–1776" (Ph.D. diss., Stanford University, 1968), p. 170, estimated a total of 565 members for New Warke and Wilmington in 1757.

21. R. V. Wells, "Quaker Marriage Patterns in a Colonial Perspective," *WMQ*, 2d ser. 29 (1972):420.

22. The best discussions of the relation of marriage to production are in Ann Kussmaul, "Time and Space, Hoofs and Grain: The Seasonality of Marriage in England," *JIH* 15 (1985):755–779; her "Agrarian Change in Seventeenth-Century England: The Economic Historian as Paleontologist," *JEH* 45 (1985):1–31; and Edward Byers, "Fertility Transition in a New England Commercial Center: Nantucket, Massachusetts, 1680–1840," *JIH* 13 (1982):17–40. No pattern for North America has yet been determined by historians. According to Daniel Scott Smith, conceptions peaked somewhat earlier in Massachusetts, in November to January, and somewhat later in England, in May to July. Communication to the author April 9, 1985.

23. Kenneth A. Lockridge, "The Population of Dedham, Massachusetts, 1636–1736," *EHR*, 2d ser. 19, no. 2 (1966):339–340; Daniel Scott Smith, "A Homeostatic Demographic Regime: Patterns in West European Family Reconstitution Studies," in *Population Patterns in the Past*, ed. Ronald L. Lee (New York: Academic Press, 1977), pp. 23, 28, 40–41; and Henri Leridon, *Natalité, saisons et conjoncture economique* (Paris: Presses Universitaires de France, 1973).

24. For Elizabeth Reed Levis, see Heiss, *Quaker Biographical Sketches*, p. 318.

25. Levy, "Light in the Valley," especially pp. 31–42, and his " 'Tender Plants': Quaker Farmers and Children in the Delaware Valley, 1681–1735," *JHF* 3 (1978):116–135.; Jack D. Marietta, "Quaker Family Education in Historical Perspective," *QH* 63 (1974):11; and his *The Reformation of American Quakerism, 1748–1783* (Philadelphia: University of Pennsylvania Press, 1984).

26. Ibid., p. 159.

27. Daniel Scott Smith, "Parental Power and Marriage Patterns: An Analysis of Historical Trends in Hingham, Massachusetts," *JMF* 35 (1973):419–428, and Daniel Scott Smith and Michael S. Hindus, "Premarital Pregnancy in America, 1640–1971: An Overview and Interpretation," *JIH* 4 (1975):537–570.

28. Levy, "Light in the Valley," p. 182.

29. Marietta, *The Reformation*, pp. 10–19.

30. Ibid., p. 71.

31. Susan Forbes, " 'As Many Candles Lighted': The New Garden Monthly Meeting, 1718–1774" (Ph.D. diss., University of Pennsylvania, 1972), p. 138; Marietta, "Ecclesiastical Discipline," p. 208, and his *The Reformation*, pp. 51, 66–67; Records of the Kennett Monthly Meeting, FHLSC; Fraizer's will is in "Abstracts of Wills, Chester County," CCHS; Levy, "Light in the Valley," p. 182, says it was uncommon for children marrying out to be disinherited.

32. Ibid.; Marietta, "Quaker Family Education," p. 15; Martin E. Lodge, "The Crisis of the Churches in the Middle Colonies, 1720–1750," *PMHB* 95 (1971):195–220.

33. Nancy Tomes, "The Quaker Connection," in *Friends and Neighbors: Group Life in America's First Plural Society*, ed. Michael Zuckerman (Philadelphia: Temple University Press, 1982), pp. 174–195.

34. Removals, 1750, Records of the Kennett Monthly Meeting, FHLSC.

35. Ibid.

36. New Castle County Tax Lists, DHR; Kennett Township Tax List, CCHS; James

T. Lemon, *The Best Poor Man's Country: A Geographical Study of Early Southeastern Pennsylvania* (Baltimore: Johns Hopkins University Press, 1972), p. 19.

37. Barry Reay, "The Social Origins of Early Quakerism," *JIH* 11 (1980):55–72; Lemon, *Best Poor Man's Country*, p. 20.

38. Shur, "Emergence of Free Schools," p. 258; Howard H. Brinton, "Friends of the Brandywine Valley," *QH* 51 (1962):67–86. Although often identified as a Quaker culture area, scholars have yet to define the content precisely.

CHAPTER 2

1. Harlan, *History and Genealogy of the Harlan Family*; Sarah Harlan Will #1089, CCCH.

2. Marylynn Salmon, "The Property Rights of Women in Early America: A Comparative Study" (Ph.D. diss., Bryn Mawr College, 1980), pp. 280–286.

3. Norma Basch, *In the Eyes of the Law: Women, Marriage and Property in Nineteenth-Century New York* (Ithaca: Cornell University Press, 1982), pp. 17–24.

4. Miranda Chaytor, "Household and Kinship: Ryton in the Late 16th and Early 17th Centuries," *History Workshop* 10 (Autumn 1980):25–60; Basch, *In the Eyes of the Law*, p. 72.

5. Ibid., p. 107; Jack Goody, "Inheritance, Property, and Women: Some Comparative Considerations," in *Family and Inheritance: Rural Society in Western Europe, 1200–1800* ed. Jack Goody, Joan Thirsk, and E. P. Thompson (Cambridge: Cambridge University Press, 1976), p. 10.

6. David Sabean, "Aspects of Kinship Behaviour and Property in Rural Western Europe before 1800," in Ibid., p. 105.

7. Salmon, "The Property Rights of Women in Early America," especially pp. 169–211, 263, 274–286; and her "Equality or Submersion? Femme Covert Status in Early Pennsylvania," in *Women of America: A History*, ed. Carol Ruth Berkin and Mary Beth Norton (Boston: Houghton Mifflin, 1979), pp. 92–113. See also Daniel Snydacker, "Kinship and Community in Rural Pennsylvania, 1749–1820," *JIH* 13 (1982):41–61.

8. Wills by women 1750s in CCCH.

9. Wills by men 1750s in CCCH. Lisa Waciega, "A 'Man of Business': The Widow of Means in Philadelphia and Chester County, 1750–1850" (unpublished paper, 1985), argues that widows were receiving more personal property from husbands by the 1790s. Her forthcoming dissertation at Temple University will provide a clearer picture of this change.

10. Wills by women 1790s in CCCH.

11. Wills by men 1790s in CCCH. Hollingsworth, "Irish Quakers," pp. 150–162; Nancy Osterud and John Fulton, "Family Limitation and Age at Marriage: Fertility Decline in Sturbridge, Massachusetts, 1730–1850," *PS* 30 (1976):481–487. For discussions of later restrictions, see Gary L. Laidig, Wayne A. Schujter, and C. Shannon Stokes, "Agricultural Variation and Human Fertility in Antebellum Pennsylvania," *JFH* 6 (1981):195–204.

12. Wells, "Quaker Marriage Patterns," p. 426.

13. Ibid., p. 438, gives age as 23.4 and size as 6.2; Levy, "Light in the Valley," p. 92, also finds Welsh Quaker family size 6, although marrying ages slightly higher.

14. Wells, "Quaker Marriage Patterns," p. 420.

15. Robert V. Wells, "Family Size and Fertility Control in Eighteenth-Century America: A Study of Quaker Families," *PS* 25 (1971):81. Byers, "Fertility Transition in a New England Commercial Center," pp. 17–40, argues that family limitation in this Quaker community had begun by the 1740s, because of the striking increase in the interval between

the penultimate and last birth. See the reply of Barbara J. Logue, "The Case for Birth Control before 1850: Nantucket Reexamined," *JIH* 15 (1985):371–391.

16. Brissot de Warville, *New Travels*, p. 172.

17. Jean-Louis Flandrin, "Contraception, Marriage, and Sexual Relations in the Christian West," in *Biology of Man in History: Selections from the Annales Économies, Sociétés Civilisations*, ed. Robert Forster and Orest Ranum, trans. Elborg Forster and Patricia M. Ranum (Baltimore and London: Johns Hopkins University Press, 1975), pp. 42–47; André Burguiére, "From Malthus to Max Weber: Belated Marriage and the Spirit of Enterprise," in *Family and Society: Selections from the Annales Economies, Sociétés, Civilisations*, ed. Robert Forster and Orest Ranum, trans. Elborg Forster and Patricia M. Ranum (Baltimore and London: Johns Hopkins University Press, 1976), p. 240; and Lawrence Stone, *The Family, Sex and Marriage in England, 1500–1800* (New York: Harper & Row, 1977), p. 417.

18. Henri Leridon, *Human Fertility: The Basic Components*, trans. Judith F. Helzner (Chicago and London: University of Chicago Press, 1977), pp. 117–120.

19. Ibid., p. 25; and Kenneth and Anna M. Roberts, eds., *Moreau de St. Méry's American Journey [1793–1798]* (Garden City, N.Y.: Doubleday, 1947), p. 315.

20. William Darlington Papers, Diary, 1822–1846, Microfilm reel 7, NYHS; Gilbert Cope, *Genealogy of the Darlington Family* (West Chester, Pa., n.p., 1900), pp. 76, 92–93; Janet McClintock Robison, "Country Doctors in the Changing World of the Nineteenth Century: An Historical Ethnolography of Medical Practice in Chester County, Pennsylvania, 1790–1861" (Ph.D. diss., University of Pennsylvania, 1975); and Larry Lee Burkhart, "The Good Fight: Medicine in Colonial Pennsylvania, 1681–1765" (Ph.D. diss., Lehigh University, 1982).

21. William Darlington, "A Medical Diary and Journal of Such Transactions, and Observations, as Are Related to Medicine," Birmingham, Pa., 1804, Ms. 3423, CCHS; "Obstetric Synopsis Containing a Regular List of Those women Who Have Been Delivered by Wm. Darlington, M.D.," Ms. 2986, CCHS.

22. Claire Elizabeth Fox, "Pregnancy, Childbirth and Early Infancy in Anglo-American Culture: 1675–1830" (Ph.D. diss., University of Pennsylvania, 1966), pp. 224–225. David Hackett Fischer, *Growing Old in America: The Bland-Lee Lectures Delivered at Clark University* (New York: Oxford University Press, 1977), p. 228, gives the estimated age of menarche as 15.2 in 1800. The trend since 1800 has been to drop about ten months every fifty years; thus the eighteenth-century age may have been near 16.

23. Darlington, "Medical Diary," 8–19–1811, 11–5–1811, 2–24–1812, CCCH.

24. Darlington, Ibid., 12–13–1809.

25. Fox, "Pregnancy," pp. 350–351.

26. Ibid., pp. 96–99; Darlington, "Medical Diary," 10–14–1808.

27. Ibid., 11–5–1811 for childbed fever.

28. Ibid., 12–5–1804.

29. Ibid., 7–24–1805; 7–30–1805.

30. Ibid.

31. Ibid.

32. Ibid., 11–1–1804.

33. Suzanne Dee Lebsock, "Women and Economics in Virginia: Petersburg, 1784–1820" (Ph.D. diss., University of Virginia, 1977), and her *The Free Women of Petersburg: Status and Culture in a Southern Town, 1784–1860* (New York: Norton, 1984), pp. 15–53; Linda E. Speth, " 'More than Her Thirds': Wives and Widows in Colonial Virginia," *Women & History*, no. 4 (1982):5–42; Susan Dwyer Amussen, "Governers and Governed: Class and Gender Relations in English Villages, 1590–1725" (Ph.D. diss., Brown University, 1982), pp. 177–181.

CHAPTER 3

1. Alexander Marshall, " 'The Days of Auld Lang Syne': Recollections of How Chester

Countians Farmed and Lived Three-Score Years Ago," ed. Don Yoder, *Pennsylvania Folklore* 13 (July 1964):13–14.

2. Ibid., p. 11. There is no careful chronology for the development of scythes or sickles. Both serrated and smooth-edged sickles were in use by the early nineteenth century. See Henry Stephens, *The Book of the Farm: Detailing the Labors of the Farmer, Steward, Plowman, Hedger, Cattle-Man, Shepherd, Field-Worker, and Dairymaid*, 2 vols. (New York: Greeley & McElrath, 1847), 2:329, who lists the smooth-edged sickle as more productive than the serrated.

3. References to women's work pulling and processing flax are frequent in the later eighteenth century, but see especially ibid., p. 14, who specifically refers to it as "women's work."

4. The farm cycle can be traced in farm journals. See especially Benjamin Hawley, Diary, 1761–1763, 1770–1775, CCHS.

5. Marshall, "Days of Auld Lang Syne," p. 15.

6. Benjamin Hawley, Diary, CCHS.

7. Marshall, "Days of Auld Lang Syne," pp. 13–14.

8. David Walter Galenson, *White Servitude in Colonial America: An Economic Analysis* (Cambridge: Cambridge University Press 1981).

9. Ibid., pp. 23–24, 226–227.

10. Sharon V. Salinger, " 'Send No More Women': Female Servants in Eighteenth-Century Philadelphia," *PMHB* 107 (1983):29–48.

11. Sharon V. Salinger, "Labor and Indentured Servants in Colonial Pennsylvania" (Ph.D. diss., University of California, Los Angeles, 1980), and her "Colonial Labor in Transition: The Decline of Indentured Servitude in Late Eighteenth Century Philadelphia," *Labor History* 22 (1981):165–191. Papers of the Pennsylvania Abolitionist Society, HSP, Microfilm reel 22, Indentures.

12. Allen Tully, "Patterns of Slaveholding in Colonial Pennsylvania: Chester and Lancaster Counties, 1729–1758," *JSH* 6 (Spring 1973):284–305; Watson, "Memorials," p. 172.

13. Deborah Nayle, Will #1389, October 5, 1750, CCCH.

14. Darold D. Wax, "Africans on the Delaware: The Pennsylvania Slave Trade, 1759–1765," *PH* 50 (1983):38–49; Merle G. Brouwer, "Marriage and Family Life among Blacks in Colonial Pennsylvania," *PMHB* 99 (1975):368–372; and Jean R. Soderlund, "Conscience, Interest, and Power: The Development of Opposition to Slavery among Quakers in the Delaware Valley, 1688–1780" (Ph.D. diss., Temple University, 1981), pp. 196–200. The number of blacks on the Chester County tax list for 1760 was 279. Of the 222 for whom an age was listed, 59 percent were between the ages of 10 and 29. Three other age categories—1–9, 30–39, 40 and over—accounted for only 14 percent each (Chester County Tax List, 1760, in the Shippen Papers, HSP).

15. Soderlund, "Conscience, Interest, and Power," p. 117 *n*128.

16. Pennsylvania Abolitionist Society, HSP, Microfilm reels 22, 23, Indentures; and Debra L. Newman, "Black Women in the Era of the American Revolution in Pennsylvania," *JNH* 61 (1976):286.

17. Lucy Simler, "The Township: The Community of the Rural Pennsylvania," *PMHB* 106 (1982):41–68.

18. Goshen Town Book, CCCH; Diary of John Sugar, 1793–1801, CCHS. For a larger view of the yearly live-in servant, see Ann Kussmaul, *Servants in Husbandry in Early Modern England* (Cambridge: Cambridge University Press, 1981).

19. Tax lists, 1767, CCCH; Forbes, "As Many Candles," p. 27.

20. John Buffington accounts, CCHS; tax lists, 1771, CCCH.

21. Benjamin Hawley, Diary, CCHS.

22. Account of Richard Barnard, CCHS.

23. Account Book of William Smedley, 1751–1766, CCHS. This interdependence is

also noted by Bettye Hobbs Pruitt for Massachusetts in "Self-Sufficiency and the Agricultural Economy of Eighteenth-Century Massachusetts," *WMQ*, 2d ser. 41 (1984):333–364. For a wide variety of work performed by rural women in Bucks County, see Joseph E. Walker, *Hopewell Villages: the Dynamics of a Nineteenth Century Iron-Making Community* (Philadelphia: University of Pennsylvania, 1974, reprint of 1966 ed.), pp. 322–324.

24. Benjamin Rush, *An Account of the Manners of the German Inhabitants*, (Philadelphia: Samuel P. Town, 1875) p. 25; John Watson, "An Account of the First Settlement of the Townships of Buckingham and Solebury in Bucks County, Pennsylvania," Historical Society of Pennsylvania, *Memoirs* 1 (1826):291; Biddle, *Journal of Elizabeth Drinker*, p. 232; Roberts, *Moreau*, pp. 84–88; account books of Martha Lewis, 1827–1835, CCHS.

25. Leo Rogin, *The Introduction of Farm Machinery in its Relation to the Productivity of Labor in the United States during the Nineteenth Century* (Publications in Economics, vol. 9 (Berkeley: University of California, 1931), p. 72; M. Roberts, "Sickles and Scythes: Women's Work and Men's Work at Harvest Time," *History Workshop* 7 (1979):3–28; K. D. M. Snell, "Agricultural Seasonal Unemployment, the Standard of Living, and Women's Work in the South and East, 1690–1860," *EHR* 34 (1981):407–433; David Tresemer, *The Scythe Book* (Brattleboro, Vt: Hand and Foot, 1981), pp. 18, 75; Peter H. Cousins, *Hog Plow and Sith: Cultural Aspects of Early Agricultural Technology* (Greenfield Village & Henry Ford Museum, no date), pp. 11, 12, 14; Axel Steensberg, *Ancient Harvesting Implements: A Study in Archeology and Human Geography* (Kobenhavn: Bianco Lunos Bogtrykker: 1943).

26. J. A. Perkins, "Harvest Technology and Labour Supply in Lincolnshire and the East Riding of Yorkshire, 1750–1850," *Tools and Tillage* 3, no. 1 (1976):54, 129, and 3, no. 2 (1977):125–135, discusses the difficulty of short Irish males in using the scythe as well.

27. Ibid.; Lemon, *Best Poor Man's Country*, p. 156.

28. Snell, "Agricultural Seasonal Employment," pp. 407–433.

29. Rogin, *Introduction of Farm Machinery*, pp. 70–72; Edwin Morris Betts, ed., *Thomas Jefferson's Farm Book* (Charlottesville: University Press of Virginia, 1953), p. 58.

30. Chester and New Castle County Wills, CCCH, DHR and CC.

31. Margaret Allinson to child, 5–21–1784, refers to keeping hens, ducks, and turkeys, Allinson Collection, QCHC; Roberts, *Moreau*, pp. 84–88.

32. Michel, "In a Manner and Fashion"; Mary Jenkins, Will #1426, CCCH; and James Maxwell Will, November 1798, DHR.

33. See particularly Lorena S. Walsh, "Urban Amenities and Rural Sufficiency: Living Standards and Consumer Behavior in the Colonial Chesapeake, 1643–1777," *JEH* 43 (1983):109–117, and Cary Carson and Lorena S. Walsh, "The Material Life of the Early American Housewife," *WP* (forthcoming).

34. Susan Burrows Swan, *Plain & Fancy: American Women and Their Needlework, 1750–1850* (New York: Holt, Rinehart & Winston, 1977), pp. 65, 143, 203.

35. Michel Foucault, *The Order of Things: An Archaeology of the Human Sciences* (New York: Random House, 1970), pp. 7–13.

36. Morris Oskidlen, Will #1648, August 15, 1755, CCCH. The looking glass trade is discussed in Alfred Coxe Prime, *The Arts & Crafts in Philadelphia, Maryland and South Carolina* 2 vols. (New York, Da Capo Press, 1969 reprint), 2:194–196.

37. John Stauffer Hartman, "The Contribution of Chester County to the Revolutionary Cause, 1774–1783" (Master's thesis, Pennsylvania State College, 1936); Anne M. Ousterhout, "Opponents of the Revolution Whose Pennsylvania Estates Were Confiscated," *Pennsylvania Genealogical Magazine* 30 (1978):237–253; and accounts of the battle by Henry Pleasants, Jr., and George B. McCormick, in *Southeastern Pennsylvania*, ed. J. Bennett Nolan (Philadelphia: Lewis Historical Publishing, 1943), pp. 207–209, 240–241.

38. Jack D. Marietta, "Wealth, War and Religion: The Perfecting of Quaker Asceticism 1740–1783," *Church History* 43 (1974):230–241.

39. Boiling tea anecdote is in Watson, "Memorials," p. 165; Marc Egnal, "The Economic Development of the Thirteen Continental Colonies, 1720 to 1775," *WMQ* 3d ser. (1975):191–222.

40. Depredation claims are in Chester County, Miscellaneous Papers, Depredation Claims, HSP.

41. Ibid.

42. Ibid.

43. Brinton, "Friends of the Brandywine Valley," p. 86.

44. Quaker Claims are in QCHC.

45. Brissot de Warville, *New Travels*, pp. 204–208, 162–164.

46. Ernest Bohn, "Geography and History of America," ms. translation by Christopher Daniel Ebeling, 1883, HSP, of 1799 work originally published in Hamburg.

47. E. P. Thompson, "Time, Work-Discipline, and Industrial Capitalism," *Past and Present* 38 (1967):56–97; Mary Douglas and Baron Isherwood, *The World of Goods: Towards An Anthropology of Consumption* (New York: Basic Books, 1979); and Prime, *Arts & Crafts*, 2:227, 273.

48. Charity Garrett, Will #4739, Alice Piersol, Will #4078, CCCH.

CHAPTER 4

1. Simler, "The Township: The Community of the Rural Pennsylvanian," pp. 49–54.

2. William Clinton Heffner, *History of Poor Relief Legislation in Pennsylvania, 1682–1913* (Cleona, Pa.: Holzapfel, 1913); Ramon Powers, "Wealth and Poverty: Economic Base, Social Structure, and Attitude in Prerevolutionary Pennsylvania, New Jersey, and Delaware" (Ph.D. diss., University of Kansas, 1971), pp. 229–335, discusses poor laws; "A List of the Names of the Annual Publick Officers in Kennett Township," HSP.

3. Gary B. Nash, "Poverty and Poor Relief in Pre-Revolutionary Philadelphia," *WMQ* 3d ser. 33 (1976):3–30, is the best discussion of urban welfare. See also John K. Alexander, *Render Them Submissive: Responses to Poverty in Philadelphia, 1760–1800* (Amherst: University of Massachusetts Press, 1980), and Simler, "The Township," p. 59.

4. James T. Mitchell and Henry Flanders, comps., *The Statutes at Large of Pennsylvania from 1682 to 1801*, 16 vols. (Philadelphia, 1896–1911), vol. 3 (1712–1724).

5. Ibid., vol. 8 (1770–1776).

6. Miscellaneous Indentures and Petitions, Poorhouse Records, CCCH.

7. Ibid.

8. Ibid.

9. West Fallowfield Town Book, 1765–1803, Ms. 13222, CCHS.

10. Ibid.

11. West Bradford Township Book, 1736–1799, CCHS.

12. William Pim, Book for Recording Assignments of Servants, 1739–1751, ms. 13521a, CCCH.

13. A Book for the Poor of the Township of East Caln, 1735–1750, CCHS; West Bradford Township Book, 1736–1799, ms. 76209, CCHS.

14. Miscellaneous Indentures and Petitions, Poorhouse Records, CCCH.

15. Ibid.; West Fallowfield Town Book, 1765–1803, ms. 13222, CCHS.

16. West Whiteland Town Book, 1765–1796, CCHS.

17. Walter I. Trattner, *From Poor Law to Welfare State: A History of Social Welfare in America* (New York: Free Press, 1974), pp. 44–73, traces the transition. Other counties trailed Chester County. Berks County, for example, did not establish its poorhouse until 1824. See Robert F. Ulle, "Blacks in Berks County, Pennsylvania: the Almshouse Records," *Pennsylvania Folklife* 27 (1977):19–30.

18. Directors of the Poor, Day Book, 1798–1824, Poorhouse Records, CCCH.
19. Record of Admission from 1800 to 1826, Poorhouse Records, CCCH.
20. Description of building is in Report of the Visitors to the Chester County Poorhouse, 1853, Poorhouse Records, CCCH.
21. Directors of the Poor, Day Book, 1789–1824, 1, 12–16, CCCH.
22. Ibid., pp. 17–24.
23. As an example of farm income, see reports by Elizabeth Yarnall, matron, and Walker Yarnall, steward, on farm produce, *American Republic*, 12–22–46.
24. David. J. Rothman, *The Discovery of the Asylum: Social Order and Disorder in the New Republic* (Boston: Little, Brown, 1971), pp. 130–154, discusses this movement.
25. Report of the Visitors to the Chester Country Poorhouse, 1845, Poorhouse Records, CCCH.
26. Rothman, *Discovery of the Asylum*, p. 154.
27. Cassy's story can be traced in Record of Admissions from 1800 to 1826, Poorhouse Records, CCCH.
28. Admissions and Examinations Book, 1841–1851, Inmates' Personal Accounts and Life Histories, Poorhouse Records, CCCH.
29. Superintendant's Reports, 1806–1842, Poorhouse Records, CCCH.
30. Return of the Poor of Chester County, 1801–1823, Poorhouse Records, CCCH; Outdoor Allowance Book, 1801–1827, Poorhouse Records, CCCH.
31. Book of Monthly Reports of the Steward, 1825–1827, Poorhouse Records, CCCH.
32. Minutes of the Directors of the Poor, 1798–1819, Poorhouse Records, CCCH.
33. Book of Monthly Reports of the Steward, 1825–1827, Poorhouse Records, CCCH.
34. Account Book with Other County Poorhouses, 1823–1850, and Report of the Visitors to the Chester County Poorhouse, 1845, Poorhouse Records, CCCH; Heffner, *History of Poor Relief*, p. 205, discusses criticism of Chester County Poorhouse in 1830s. A similar rise in the numbers of dependent poor women occurred in rural New York according to Joan Underhill Hannon, "Poverty in the Antebellum Northeast: The View from New York State's Poor Relief Rolls," *JEH* 44 (1984):1014.
35. Oblinger, "Alms for Oblivion," pp. 94–119.
36. Negley K. Teeters, *Scaffold and Chair: A Compilation of Their Use in Pennsylvania, 1682–1962* (Philadelphia: Pennsylvania Prison Society, 1963), p. 20–21; Negley K. Teeters, "Tentative List of Hangings in County Jails of Pennsylvania, 1834–1913" (Temple University, Spring 1961, Mimeographed), pp. 8, 23, CCHS; Sharon Ann Burnston, "Babies in the Well: An Underground Insight into Deviant Behavior in Eighteenth-Century Philadelphia," *PMHB* 106 (1982):151–186. Dr. William Darlington reported Hannah Miller's public execution on August 1, 1805, as a "dismal sight," "Medical Diary," CCHS. Wilson's fate was recounted in narrative form and read by Quaker women like Elizabeth Drinker in 1797, Biddle, *Journal of Elizabeth Drinker*, p. 303.
37. Dockets, Gaol Keepers, 1804–1807, 1812–1816, CCCH.
38. Ibid.
39. Public Offices, Prison, Tabular View of the Convict Department of the Chester County Prison from August 6, 1840, CCHS.
40. Richard A. Cloward and Frances Fox Piven, "Hidden Protest: The Channeling of Female Innovation and Resistance," *Signs* 4 (1979):651–669.
41. Chester County Poorhouse, Record Books of Diseases Treated, 1841–1842, CCCH. This may have been a problem of diagnosis. Ellen Dwyer reported no diagnosis of syphilitic women in the Utica, New York, mental asylum although 9 percent of the men had it, "Sex Roles and Psychopathology: A Historical Perspective," in *Sex Roles and Psychopathology*, ed. Cathy S. Widom (New York: Plenum Press, 1984).
42. Ibid.
43. Harold B. Hancock, "The Indenture System in Delaware, 1681–1921," *DH* (1974–1975):47–59; Chester County Indentures, "Record of the Names of Children Bound to

Trades or Service," CCCH; New Castle County Indentures, DHR.

44. Ibid.

45. Ibid.

46. Ibid.

47. Ibid.

48. Description of conditions for blacks is in Report of the Visitors to the Chester County Poorhouse, Poorhouse Records, CCCH.

49. Infirmary and insane in Visitors Report to the Chester County Poorhouse, 1841, Poorhouse Records, CCCH.

50. Report of the Visitors to the Chester County Poorhouse, 1853, Poorhouse Records, CCCH. The argument that only curables were transferred is in Rothman, *Discovery of the Asylum*, p. 131. Nancy Tomes, "The Domesticated Madman: Changing Concepts of Insanity at the Pennsylvania Hospital, 1780–1830," *PMHB* 106 (1982):278, argues that only the most dangerous lunatics were put in the earlier hospital for the insane. In New York, the state asylum also returned patients to the county asylums if they did not respond to treatment after two years. The majority of patients were women. Ellen Dwyer, "The Weaker Vessel: Legal Versus Social Reality in Mental Commitments in Nineteenth Century New York," in *Women and the Law: Social Historical Perspectives*, ed. Kelly Weisberg (Cambridge, Mass.: Schenkman, 1983), pp. 87, 91.

51. Report of the Visitors to the Chester County Poorhouse, 1853, Poorhouse Records, CCCH. Buildings discussed in Francis James Dallett, *The Athenaeum Age in Chester County* (West Chester: Chester County Historical Society, 1961), pp. 21–22. Gerald N. Grob, "Class, Ethnicity, and Race in American Mental Hospitals, 1830–1875," in *Theory and Practice in American Medicine*, ed. Gert H. Briegen, (New York: Science History Publications, 1976), p. 242, points out the lack of care for blacks. Norman Dain, *Concepts of Insanity in the United States, 1789–1865* (New Brunswick: Rutgers University Press, 1964), pp. 22–35, discusses the Quaker asylum.

52. Ledger and Boarders Record, Trustees of the Poor for New Castle County, vol. 1, 1822–1832, vol. 2, 1833–1850, DHR.

53. Cloward and Piven, "Hidden Protest," pp. 651–669.

54. Michael MacDonald, *Mystical Bedlam: Madness, Anxiety, and Healing in Seventeenth-Century England* (Cambridge: Cambridge University Press, 1981), pp. 36–41, 73.

55. The increase in commitments in New York is also noted in Dwyer, "Sex Roles and Psychopathology," pp. 31–33.

56. Sandra M. Gilbert and Susan Gubar, *The Madwoman in the Attic: The Woman Writer and the Nineteenth Century Literary Imagination* (New Haven and London: Yale University Press, 1979), pp. 51, 86; Emily Dickinson quote in Adrienne Rich, *On Lies, Secrets, and Silence: Selected Prose, 1966–1978* (New York: Norton, 1979), p. 175.

CHAPTER 5

1. Lemon, *Best Poor Man's Country*, pp. 218–228; James T. Lemon and Gary B. Nash, "The Distribution of Wealth in Eighteenth Century America: A Century of Change in Chester County, Pennsylvania, 1693–1808," *JSH* 2 (1968):1–24; Alan Tully, "Economic Opportunity in Mid-Eighteenth Century Rural Pennsylvania," *Social History/Histoire Sociale* 9 (1976):111–128; William S. Sachs, "Agricultural Conditions in the Northern Colonies before the Revolution," *JEH* 13 (1953):274–290; D. E. Ball and G. M. Walton, "Agricultural Productivity Change in Eighteenth Century Pennsylvania," *JEH* 36 (1976):102–117; Diane Lindstrom, "American Economic Growth before 1840: New Evidence and New Directions," *JEH* 39 (1979):289–301; and her "The Industrialization of the East, 1810–1860," *Working Papers from the Regional Economic History Research Center* 2, no. 3

(1979):19–37; her *Economic Development in the Philadelphia Region, 1810–1850* (New York: Columbia University Press, 1978); David M. Gordon, Richard Edwards, and Michael Reich, *Segmented Work, Divided Workers: The Historical Transformation of Labor in the United States* (Cambridge: Cambridge University Press, 1982), pp. 54–68.

2. For export and coastal trade, see Oakes, "A Ticklish Business," pp. 209–211; Lindstrom, "Industrialization of the East," p. 19, and her *Economic Development*, p. 194; and James F. Shepherd and Gary M. Walton, *Shipping, Maritime Trade, and the Economic Development of Colonial North America* (Cambridge: Cambridge University Press, 1972), pp. 110–112, especially 112n1. Export statistics in Adam Seybert, *Statistical Annals: Embracing Views of the Population* (Philadelphia: Thomas Dobson, 1818), p. 154.

3. Oakes, "A Ticklish Business," p. 211. While the overall influence of the export trade is still being evaluated, it seems to have been crucial in this case. See Caludia D. Golden and Frank D. Lewis, "The Role of Exports in American Economic Growth during the Napoleonic Wars, 1793 to 1807," *Explorations in Economic History* 17 (1980):6–25.

4. For Philadelphia cows and consumption, see Billy G. Smith, "The Material Lives of Laboring Philadelphians, 1750 to 1800," *WMQ*, 3d ser. 33 (1981):170, 174.

5. Gary B. Nash and Billy G. Smith, "The Population of Eighteenth-Century Philadelphia," *PMBH* 99 (1975):366, and Billy G. Smith, "Death and Life in a Colonial Immigrant City: A Demographic Analysis of Philadelphia," *JEH* 37 (1977):865. Twentieth-century consumption estimates in E. G. Montgomery and C. H. Kardell, *Apparent Per Capita Consumption of Principal Foodstuffs in the United States* (Washington, 1930), p. 25; and in Richard Osborn Cummings, *The American and His Food* (Chicago: University of Chicago Press, 1940), p. 236. But see also Cummings, pp. 241–243, for earlier estimates for 1833, 1851, and 1863, and X. A. Willard, *Practical Butter Book* (New York: Rural Publishing 1875), p. 6.

6. Woodward, *Ploughs and Politics*, pp. 343–344. Diary of Benjamin Hawley, CCHS.

7. Richard Buffington, Arithmetic Book, 1770, CCHS.

8. Anne Bezanson, Robert D. Gray, and Miriam Hussey, *Prices in Colonial Pennsylvania* (Philadelphia: University of Pennsylvania Press, 1935), pp. 62, 316–317.

9. Claims are in Chester County, Miscellaneous Papers, Depredation Claims, HSP. Quaker accounts are at QCHC, in the Meeting for Sufferings papers. There was little butter-making equipment listed in the Quaker accounts but many cows and heifers. Biddle, *Journal of Elizabeth Drinker*, pp. 62–64, 106; Nicholas Wainwright, ed., "A Diary of Trifling Occurrences: Philadelphia, 1778," *PMHB* 82 (1958):457; Anne Bezanson, *Prices and Inflation during the American Revolution: Pennsylvania, 1770–1790* (Philadelphia: University of Pennsylvania Press, 1951), p. 315–316.

10. Samuel Taylor Journal, 1798–1799, CCHS.

11. Special Accounts, 1803–1822, Series C, Mme. E. I. du Pont, Box 14; Group 3, Eleuthère Irénée du Pont, Longwood Manuscripts, EMHL. Petitions cited in Roy M. Bootman, "The Brandywine Cotton Industry, 1795–1865," Hagley Museum, Research Report, 1957, EMHL. The bartering of butter by women is also discussed in Walker, *Hopewell Villages*, pp. 130, 200.

12. Rev. Patrick Kenney Papers, Microfilm at EMHL.

13. Anne Bezanson, Robert D. Gray, Miriam Hussey, *Wholesale Prices in Philadelphia, 1784–1861* (Philadelphia: University of Pennsylvania Press, 1937), pp. 25, 62, 99, 405.

14. Retail prices are from Hawley, Taylor, du Pont, and Kenny diaries cited above and in Eli W. Strawn Collection, Butter and Egg Book, FLMM.

15. W. Emerson Wilson, *Plantation Life at Rose Hill: The Diaries of Martha Ogle Forman, 1814–1845* (Wilmington: Historical Society of Delaware, 1976), pp. v–viii, has a brief biography of Forman; J. Smith Futhey and Gilbert Cope, *History of Chester County* (Philadelphia: Louis H. Everts, 1881), pp. 426–430, discussed Lewis.

16. Wilson, *Plantation Life*, pp. 181, 242.

17. Diaries of Esther Lewis, Lewis Papers, FHLSC.

18. Eli W. Strawn Collection, Butter and Egg Book, FLMM. Population and Agricultural Census, 1850, showed 600 pounds of butter produced in the year ending June 1, 1850.

19. U.S., Commissioner of Patents, *Annual Report, 1847* (Washington, 1848), pp. 206, 215, 364–366, 579–585.

20. Samuel Bulkley Ruggles, *Tabular Statements from 1840 to 1870 of the Agricultural Products of the States and Territories of the United States of America* (New York: Chamber of Commerce, 1874), pp. 11, 38; Lindstrom, *Economic Development*, p. 142; U.S., Secretary of the Interior, *Agriculture of the United States in 1860* (Washington, 1864), p. xxxiii. By 1860, the four states produced over 170 million pounds. Butter production in the United States increased about 50 percent during the decade.

21. Fred Bateman, "The 'Marketable Surplus' in Northern Dairy Farming: New Evidence by Size of Farm in 1860," *Agricultural History* 52 (1978):345–363.

22. John Spurrier, *The Practical Farmer: Being a New and Compendious System of Husbandry* (Wilmington: Brynberg & Andres, 1793), pp. 308–313; U.S., Commissioner of Patents, *Report, 1848* (Washington, 1849), Pt. 2, pp. 115–121; Carol Hoffacker, ed.,"Diaries of Edmund Canby, A Quaker Miller, 1822–1848," *DH* 16 (1974–1975): 131.

23. U.S., Census of Agriculture, 1850. The data set for this census was kindly supplied by Jack Michel.

24. Tenche Coxe, *A Statement of the Arts and Manufactures of the United States of America for the Year 1810* (Philadelphia: Cornman, 1814), pp. 44–45.

25. *Scribner's Monthly* 8 (1874):704–705.

26. Diaries of Esther Lewis, Lewis Papers, FHLSC.

27. Esther Lewis to Jonathan and Ann Thomas, 8–26–1837 and to Graceanne Lewis, 5–15–1842, Lewis Papers, FHLSC.

28. Esther Lewis to Rebecca and Edwin Fussell, 11–14–1841, Lewis Papers, FHLSC.

29. Susan Lukens, *Gleanings at Seventy-Five* (Philadelphia: Longstreth, 1883), pp. 66–67. Salinger, "Colonial Labor in Transition," pp. 165–191.

30. *Pennsylvania Gazette*, 6–14–1770, 6–21–1770, 7–5–1770, 5–7–1783, 1–28–1784.

31. Claudia Goldin and Kenneth Sokoloff, "Women, Children, and Industrialization in the Early Republic: Evidence from the Manufacturing Censuses," *JEH* 42 (1982):741–774, document the increase in women's wages between 1793 and 1833 but they do not convincingly show that there was no demand for women in agricultural labor, especially in the Mid-Atlantic. Donald R. Adams, Jr., "Workers on the Brandywine: The Response to Early Industrialization," *Working Papers from the Regional Economic History Research Center* 3, no. 4 (1980):7, 29, and his "The Standard of Living during American Industrialization: Evidence from the Brandywine Region, 1800–1860," *JEH* 42 (1982):903–917.

32. Esther Lewis to Edwin & Rebecca Fussell, 4–10–1839, Lewis Papers, FHLSC; and Cynthia A. Leo, "A Folk History of My Family," (Honor's Paper, Bishop Mills Historical Institute).

33. Wilson, *Plantation Life*, p. 421.

CHAPTER 6

1. Edwin Pessen, "How Different from Each Other Were the Antebellum North and South?" *AHR* 85 (1980):1119–1149; Fernand Braudel, *Afterthoughts on Material Civilization and Capitalism* (Baltimore: Johns Hopkins University Press, 1977), translated by Patricia Ranum, as well as his *Capitalism and Material Life, 1400–1800* (New York: Harper & Row, 1975) and *The Wheels of Commerce: Civilization and Capitalism, 15th–18th Century* (New York: Harper & Row, 1982).

2. Labor is discussed in Soderlund, "Black Women in Colonial Pennsylvania," pp. 57–

58, and "Conscience, Interest, and Power"; Allen Tully, "Patterns of Slaveholding in Colonial Pennsylvania: Chester and Lancaster Counties, 1729–1758." *JSH* 6 (1973):284–305; Salinger, "Colonial Labor in Transition," pp. 165–191, and "Labor and Indentured Servants in Colonial Pennsylvania." The use of mechanical reapers in the 1850s caused the next major change in harvest labor, but women in many areas continued to help with fieldwork depending on the farm family cycle and economic conditions. Alan L. Olmstead, "The Mechanization of Reaping and Mowing in American Agriculture, 1833–1870," *JEH* 35 (1975):327–352.

3. Introduction to M. Francis Guenon, *Treatise on Milch Cows*, trans. John S. Skinner (New York, 1865, reprint of 1856 ed.), p. 22. There is as yet no thorough study of the gender division of labor in dairying. Apparently, New England men moved into dairying quite early in the nineteenth century, but the division was more mixed in the Mid-Atlantic. Nancy Grey Osterud has found in her study of Broome County, New York, "Strategies of Mutuality: Relations among Women and Men in an Agricultural Community" (Ph.D. diss., Brown University, 1984), pp. 361–401, that men and women shared dairy tasks on occasion but that women maintained primary responsibility for butter making until the 1890s when creameries were introduced. Men seem to have performed the labor in creameries, and their appearance in an area may have signaled the transition. There is no mention of men doing any milking or butter processing in this region before 1850 in the extensive printed and manuscript literature I have consulted. In the 1867 report cited below, women were explicitly mentioned as still doing milking and butter making. U.S., Commissioner of Agriculture, "Philadelphia Butter," *Report, 1867* (Washington, 1868), p. 294.

4. Irrigation mentioned in Lemon, *Best Poor Man's Country*, pp. 175–176; winter fodder in Robert Sutcliffe, *Travels in Some Parts of North America in the Years 1804, 1805, and 1806* (Philadelphia: Kite, 1812), p. 206; barns in Margaret Berwind Schiffer, *Survey of Chester County, Pennsylvania Architecture: 17th, 18th and 19th Centuries* (Exton, Pa.: Schiffer, 1976), and Futhey and Cope, *History of Chester County*, p. 339. Oakes, "A Ticklish Business," p. 198, also discusses grasses. John Watson, "An Account of the First Settlement of the Townships of Buckingham and Salisbury in Bucks County, Pennsylvania (1804)," *Historical Society of Pennsylvania* 1 (1826):303, discusses lack of winter feed for cows. Rush, *An Account of the Manners of the German Inhabitants*, p. 15.

5. U.S., Direct Tax, 1798. Lists are incomplete but 41 percent of Sadsbury Township farms and 62 percent of West Caln farms had barns listed.

6. *Farmers' Cabinet* 1 (October 1, 1836) reported the first cattle show in the area. David Lloyd, *Economy of Agriculture: Being a Series of Compendious Essays on Different Branches of Farming* (Germantown, Pa.: Freas, 1832), p. 103, recommended Durhams.

7. C. E. Fussell, *The English Dairy Farmer, 1500–1900* (London, 1966), p. 254; *American Agriculturist* 4 (1845):336.

8. Fred Bateman, "Labor Inputs and Productivity in American Dairy Agriculture, 1850–1910," *JEH* 29 (1969):206–229. Contemporary discussions of milking are in James Cutbush, *The American Artist's Manual, or Dictionary of Practical Knowledge*, vol. 1, "Butter" (Philadelphia: Johnson, Warner & Fisher, 1814), and *Farmers, Mechanics, Manufacturers, and Sportsmans Magazine* 1 (May 1826):81. See also Oakes, "A Ticklish Business," p. 202, and Fussell, *English Dairy*, p. 163.

9. David L. Pallett, "Dairy Farming in Chester County Pennsylvania, 1770–1780: A Theoretical Model of the Joseph Pratt Dairy Operation for the Colonial Pennsylvania Farm Project," (Master's Paper, Pennsylvania State University, 1974), pp. 7, 10, 17.

10. Henry Glassie, *Pattern in the Material Folk Culture of the Eastern United States* (Philadelphia: University of Pennsylvania Press, 1968), pp. 8–10. U.S., Direct Tax, 1798. List for West Caln, Edgemont, Concord, and Bradford townships. Bradford had 154 square feet average, Concord 186 square feet.

11. Willard, *Practical Butter Book*, has the best brief description of the later interiors. See also Amos Long, Jr., *The Pennsylvania German Family Plan* (Breinigsville: Pennsylvania German Society, 1972), pp. 106–107.

12. Pallett, "Dairy Farming," p. 10; Esther Lewis to Jonathan and Ann Thomas, 8-26-1837, Lewis Papers, FHLSC; Willard, *Practical Butter*, p. 68. The Hatton springhouse was field measured.

13. For size of earthenware, see Arthur E. James, *The Potters and Potteries of Chester County, Pennsylvania* (Exton, Pa.: Schiffer, 1978), pp. 84–86.

14. *The Progress of the Dairy: Descriptive of the Methods of Making Butter and Cheese for the Information of Youth* (New York: Samuel Wood & Sons, 1819), p. 9; William Drowne, *Compendium of Agriculture* (Providence: Field & Maxcy, 1824), p. 177; John Bordley, *Essays and Notes on Husbandry and Rural Affairs* (Philadelphia: Dobson, 1801), p. 456; *Massachusetts Society for Promoting Agriculture* 2 (1809):69; William Townsend, *The Dairyman's Manual* (Vergennes, Vt.: n.p., 1839), p. 25; Mrs. Loudon, *The Lady's Country Companion; or, How to Enjoy a Country Life Rationally* (London: Longman, Brown, Green, & Longman, 1845), p. 105; James, *Potters and Pottery*, p. 17. Susan H. Myers to author, July 24, 1981, noted the use of stoneware for dairy-related pottery in an early nineteenth-century account from Maryland. Prime, *Arts & Crafts*, 1:126–127, for the 1785 warning.

15. Diary of Esther Lewis, December 7, 1830, Lewis Papers, FHLSC. Pans and skimmers are discussed in Cutbush, *American Artist's Manual*, vol. 1, "Dairy" and "Butter;" *Farmers, Mechanics, Manufacturers, and Sportsmans Magazine* 1 (May 1826):81; and *The Progress of the Dairy*, p. 9.

16. *Farmers, Mechanics, Manufacturers, and Sportsmans Magazine* 1 (May 1826):81, discussed temperature of 45 degrees but most later authorities give 50–55 degrees as the best temperature. See Loudon, *Lady's Country Companion*, p. 104; *Farmers' Cabinet* 1 (1837):105; Townsend, *Dairyman's Manual*, p. 19.

17. English butter markets of the sixteenth and seventeenth centuries are discussed in Eric Kerridge, *The Agricultural Revolution* (New York: Kelley, 1968), pp. 117–120, and Fussell, *English Dairy*. For inventories, see Ruth Matzkin, "Inventories of Estates in Philadelphia County, 1682–1710," (Master's Thesis, University of Delaware, 1959), p. 136, and Stephanie Grauman Wolf, *Urban Village: Population, Community, and Family Structure in Germantown, Pennsylvania, 1683–1800* (Princeton: Princeton University Press, 1976), p. 46. The Index of American Design, Catalog of Domestic Utensils, National Gallery, Washington, D.C., has only two verified eighteenth-century butter churns, both from New Jersey. The 1759–1765 churn is 29½ inches by 11½ inches. Frank Gilbert Dollear, "Butter," *McGraw-Hill Encyclopedia of Science and Technology*, 15 vols. (New York, McGraw-Hill, 1960), 2:379.

18. Examples of dash churns are in Elizabeth A. Powell, *Pennsylvania Butter Tools and Processes* (Doylestown, Pa.: Bucks County Historical Society, 1974), pp. 14–15.

19. Catalog for James S. Barron Co., *Woodenware* (New York, 1899), WML.

20. Disadvantages in Townsend, *Dairyman's Manual*, p. 112. Cuthbert W. Johnson, *The Farmer's Cyclopedia and Dictionary of Rural Affairs* (Philadelphia: Carey & Hart, 1850), p. 240, and Charles L. Flint, *Milch Cows and Dairy Farming* (Boston: Tilton, 1868), p. 226, mention its common use through the 1860s. For similar advice, see Cutbush, *American Artist's Manual*, vol. 1, "Churns"; Thomas Green Fessenden, *The Husbandman and Housewife: A Collection of Valuable Recipes and Directions, Relating to Agriculture and Domestic Economy* (Bellows Falls, Vt.: Bill, Blake, 1820), p. 19; Drowne, *Compendium of Agriculture*, p. 178; and *The Farmers, Mechanics, Manufacturers, and Sportsmans Magazine* 1 (May 1826):122.

21. Cutbush, *American Artist's Manual*, vol. 1, "Butter."

22. Pallett, "Dairy Farming," p. 46. The estate of Charles Read IV in New Jersey

showed one barrel churn and frame in December 6, 1783. He had seven milk cows and his father conducted early dairy experiments, Carl R. Woodward, *Ploughs and Politics: Charles Reed of New Jersey and His Notes on Agriculture, 1715–1774* (New Brunswick, N.J., Princeton University Press, 1941), p. 407. Illustrations are in *Progress of the Dairy*, p. 10. Fussell, *English Dairy*, p. 166, notes the first patent for a barrel churn in 1777 but some in use previously.

23. Disadvantages are in *Massachusetts Society for Promoting Agriculture* 2 (1806):77; *Farmers, Mechanics, Manufacturers and Sportsmans Magazine* 1 (June 1826):122, for churn strokes, and Stephans, *The Book of the Farm*, 2:307. Guenon, *Treatise on Milch Cows*, p. 25, also gave one and a half hours as the time to churn in 1846.

24. Flint, *Milch Cows*, pp. 227–229.

25. The total patents for butter-related machinery are from M. D. Leggett, *Subject-Matter Index of Patents for Inventions Issued by the U.S. Patent Office, 1790 to 1873*, 2 vols. (Washington, 1874), 1:184–186, 294–309, 246.

26. *Franklin Journal* (1832), 1:63, 85; (1833), 1:20; (1836), 1:129.

27. Figures for the location of patentees are: New York, 90; New England, 79; Ohio and Pennsylvania, 16. There are a few from Missouri, New Jersey, Kentucky, Virginia, and Indiana. For specific attempts to solve these problems, see the following issues of the *Franklin Journal*: Oliver Wyman's churn (1836), 1:125 for boxes; Clifton C. Stearns (1836), 1:206, and Joshua Pike (1839), 1:186, for centering butter; Charles Otis (1836), 1:62, and Allen and William A. Crowell (1841), 2:131 for temperature; Lewis Hinkson for foot operation (1835), 2:84; Ira Park (1835), 2:92, for washing and churning; Nathan Chapen (1849), 1:307 for time.

28. Stephens, *Book of the Farm*, p. 311; Barnam, *Family Receipts*, p. 265.

29. U.S., Patent Office, *Women Inventors to Whom Patents Have Been Granted by the United States Government, 1790 to July 1, 1888* (Washington, 1888), p. 3. Information on Lettie Smith is from Josiah B. Smith, *Historical Collections of Persons, Land, Business and Events in Newton*, 4 vols. (Doylestown, Pa.: Bucks County Historical Society, 1942), typescript in the FLMM, and from the 1850 census for Upper Makefield, Bucks County. Her patent model and publicity papers are reproduced in Powell, *Pennsylvania Butter Tools*, p. 27. The *Scientific American* reported her invention in 6 (December 21, 1850):106.

30. Churning matches in *Scientific American* 4 (October 7, 1848):20, and *The Plough, the Loom, and the Anvil* 1 (February 1849): 593.

31. Oakes, "A Ticklish Business," p. 24, quotes early caveat; Stephens, *Book of the Farm*, 2:311; H. L. Barnam, *Family Receipts or Practical Guide for the Husbandman and Housewife* (Cincinnati, Ohio: Lincoln, 1831), p. 265, has another early warning. Marble and other solutions are in Bordley, *Essay and Notes*, p. 274; Cutbush, *American Artist's Manual* vol. 1, "Butter"; *Farmers, Mechanics, Manufacturers, and Sportsmans Magazine* 1 (May 1826):81; and Barnam, *Family Receipts*, p. 265.

32. Acquisition notes, Mercer Museum, Doylestown, Pennsylvania.

33. Salting and preserving are discussed in Cutbush, *American Artist's Manual*, vol. 1, "Butter"; Bordley, *Essays and Notes*, p. 272; *Massachusetts Society for Promoting Agriculture* 2 (1806):77; and Roberts, *Pennsylvania Farmer*, p. 174. There was almost no discussion of the effect of potassium nitrates on the consumer and no medical interest in the question until much later, see Edward T. Reichert, "On the Physiological Action of Potassium Nitrate," *American Journal of the Medical Sciences*, n.s. 80 (1880):158–181, who reported its use for suppressing epileptic symptoms but raised no question about it as a food additive.

34. *Progress of the Dairy*, p. 15; Wilson, *Plantation Life*, p. 280; Esther Lewis Diary, 7–4–1831, first mentions printing, Lewis Papers, FHLSC; 1828 advertisement quoted in Powell, *Pennsylvania Butter*, p. 18; Barnam, *Family Receipts*, p. 265. Esther Lewis refers to butter pots throughout her 1830–1831 diary, Lewis Papers, FHLSC. Use of ice is

mentioned in Willard, *Practical Butter Book*, p. 69, 71, and in Powell, *Pennsylvania Butter*, p. 24.

35. James, *Potters and Pottery*, p. 95; Wilson, *Plantation Life*, pp. 52, 68, mentions crocks holding 25 and 35 pounds. Advice on packing is in *Progress of the Dairy*, p. 15; Cutbush, *American Artist's Manual*, vol. 1 "Butter"; Thomas Cooper, *A Treatise on Domestic Medicine* (Reading, Pa.: Getz, 1824); Drowne, *Compendium of Agriculture*, p. 177; *Farmers, Mechanics, Manufacturers, and Sportsmans Magazine* 1 (June 1826):123; and Barnam, *Family Receipts*, p. 256. Brissot de Warville, *New Travels*, pp. 199–201.

36. These conclusions from various references in Forman and Lewis diaries. Esther Lewis diaries, January 1830, refers to "Pedler women." Her account carries notations of bartering, buying at the door, from individuals, and from Philadelphia, Lewis Papers, FHLSC.

CHAPTER 7

1. The most helpful models for analyzing the complex interaction among folklore, popular culture, and literature were William R. Boscom, "Four Functions of Folklore," *JAF* 67, no. 266 (1954):333–349; Robert A. Schwegler, "Oral Tradition and Print: Domestic Performance in Renaissance England," *JAF* 93, no. 370 (1980):435–441; Joseph J. Arpad, "Between Folklore and Literature: Popular Culture as Anomaly," *Journal of Popular Culture* 9 (1975):403–422; John Michael Vlach, "Quaker Tradition and the Paintings of Edward Hicks: A Strategy for the Study of Folk Art," *JAF* 94, no. 372 (1981):145–165; and Jay Mechling, "Advice to Historians on Advice to Mothers," *JSH* 9 (1975):44–63.

2. Mary Beth Norton, "Eighteenth-Century American Women in Peace and War: The Case of the Loyalists," *WMQ*, 3d ser. 33 (July 1976):386–409; "Nineteenth-Century America: The Paradox of 'Women's Sphere,'" in *Women of America: A History*, Carol Ruth Berkin and Mary Beth Norton (Boston: Houghton Mifflin, 1979); *Liberty's Daughters: The Revolutionary Experience of American Women, 1750–1800*, (Boston and Toronto: Little, Brown, 1980); and "The Evolution of White Women's Experience in Early America," *AHR* 89 (1984):616. Laurel Thatcher Ulrich, *Good Wives: Image and Reality in the Lives of Women in Northern New England, 1650–1750* (New York: Knopf, 1982), pp. 202–214, discusses captives as public heroines.

3. Jack Zipes, *Breaking the Magic Spell: Radical Theories of Folk & Fairy Tales* (Austin: University of Texas Press, 1979).

4. Richard Bauman, "Quaker Folk-Linguistics and Folklore," in *Folklore: Performance and Communication*, ed. Dan Ben-Amos and Kenneth S. Goldstein (The Hague: Mouton, 1975), pp. 255–263. For other groups, see MacEdward Leach and Henry Glassie, *A Guide for Collectors of Oral Traditions and Folk Cultural Material in Pennsylvania* (Harrisburg: Pennsylvania Historical and Museum Commission, 1973).

5. Milton Drake, *Almanacs of the United States* (New York: Scarecrow, 1962). Almanacs used in this sample are at the EMHL and in the Painter Collection, FHLSC, Series 8, Box 23.

6. Marion Barber Stowell, *Early American Almanacs: The Colonial Weekday Bible* (New York: Burt Franklin, 1977), p. x, 122–124. The content of Franklin's almanacs seem to be directed at an urban male audience, however. They are quite different from the later almanacs. See *The Complete Poor Richard Almanacks*, 2 vols. (Barre, Mass.: Imprint Society, 1970).

7. John Mack Faragher, *Women and Men on the Overland Trail* (New Haven and London: Yale University Press, 1979), pp. 148–148, found similar gender conflict in midwestern ballads. Rev. John Taylor, *Cramer's Pittsburg Almanack* (Pittsburgh: Cramer,

Spear & Eichbaum, 1813), reproduced "The Whole Duty of Woman," written by a "Lady," which included piety in a standard repertory of womanly duties. No other almanac so much as hinted that piety should be a female virtue.

8. Kay Stone, "The Things Walt Disney Never Told Us," in *Women and Folklore*, ed. Claire R. Farrer (Austin and London: University of Texas Press, 1975), pp. 42–50. See the 1811 almanac, no title page, EMHL, "The Washing Day," for reference to children hearing stories from the female servants, "or thrilling tale / Of ghost, or witch, or murder."

9. *Poor Robin's Almanac* (Philadelphia: D. Dickinson, 1811).

10. *The Columbian Almanac* (Wilmington: Peter Brynberg, 1801). "Miss Bailey" and "Kitty Maggs and Jolter Giles" are from an 1812 almanac that had no front page; *The Pennsylvania, New Jersey, Delaware, Maryland and Virginia Almanac* (1799).

11. *The New St. Tammany Almanac* (Philadelphia: George W. Mentz, 1821); *Pennsylvania Almanac* (Philadelphia: McCarty & Davis & G. W. Mentz, 1825); *Poor Richard Improved* (Philadelphia, 1775); *The New-Jersey and Pennsylvania Almanac* (Trenton, N.J., 1813).

12. *Bioren's Town and Country Almanack* (Philadelphia: John Bioren, 1813); *Bioren's Town and Country Almanack* (Philadelphia: John Bioren, 1815); *The North American Calender, or the Columbian Almanac* (Wilmington: Porter & Sons, 1829); *Griggs's City and Country Almanack* (Philadelphia: John Grigg, 1831); and *The Delaware & Maryland Farmer's Almanac* (Philadelphia: J. M'Dowell, 1842).

13. *Poor Richard Improved* (Philadelphia, 1793); *Father Abraham's Almanack* (Philadelphia, 1794); *Father Abraham's Almanack* (Philadelphia, 1781); *The Pennsylvania, New-Jersey, Delaware, Maryland and Virginia Almanac* (1802); *Poor Richard Improved Almanack* (1811); *The New St. Tammany Almanac* (Philadelphia: George W. Mentz, 1818); *Columbian Almanac* (Philadelphia: Jos. McDowell, 1837).

14. *Father Abraham's Almanac* (Philadelphia, 1794); and *Poor Robin's Almanac* (Philadelphia: D. Dickinson, 1823).

15. *The LaFayette Almanac* (Philadelphia: G. W. Mentz, 1826).

16. *Farmers & Mechanics Almanack* (Philadelphia: Marot & Walter, 1827).

17. *Poor Robin's Almanac* (Philadelphia: D. Hogan, 1814).

18. *The New Jersey Almanack* (Trenton, 1788); *Poor Richard Improved* (Philadelphia, 1793); *Pennsylvania Almanac* (Philadelphia: McCarty & Davis, 1830).

19. *Uncle Sam's Large Almanac* (Philadelphia: Denny & Walker, 1834); *Agricultural Almanac* (Philadelphia: Prouty, Libby, & Prouty, 1840); *Rapp's House-Keepers' and Farmers' Temperance Almanack* (Thomas T. Mahan, 1844).

20. Changes in definition can be traced in the *Oxford English Dictionary* and Sir William A. Craigie and James R. Hulbert, eds., *A Dictionary of American English on Historical Principles*, 4 vols. (Chicago: University of Chicago Press, 1940), 2:794.

21. The best introduction to the agricultural development literature is Raymond Williams, *The Country and the City* (New York: Oxford, 1973), pp. 60–67.

22. J. Twamley, *Dairying Exemplified, or the Business of Cheese-Making: Laid Down from Approved Rules, collected from the most experienced dairy women, of several counties* (Warwick: J. Scharp, 1784): John Bordley, *Essays and Notes on Husbandry and Rural Affairs*, 2d ed. (Philadelphia: Dobson, 1801).

23. Thomas Cooper, *Treatise on Domestic Medicine* (Reading, Pa.: Getz, 1824).

24. E. P. Thompson, *The Making of the English Working Class*, (New York: Vintage, 1963) pp. 746–762.

25. Williams, *Country and the City*, pp. 108–119.

26. William Cobbett, *Cottage Economy* (New York, 1824).

27. T. D. Seymour Bassett, "Hannah Barnard," in *Notable American Women, 1607–1950: A Biographical Dictionary*, ed. Edward T. James, 2 vols. (Cambridge, Mass.: Harvard University Press, 1971), 1:88–90; Hannah Barnard to William Matthews, 9–6–1807, QCHC; *Dictionary of Quaker Biography*, QCHC.

28. *A Narrative of the Proceedings in America of the Society Called Quakers, in the Case of Hannah Barnard* (London: Stower, 1804).

29. Hannah Barnard, *Dialogues on Domestic and Rural Economy* (Hudson: n.p., 1820).

30. Ibid., p. 39.

31. Ibid., p. 56.

32. Ibid., pp. 13–14, 67.

33. Ibid., p. 53.

34. L. Davidoff, "The Role of Gender in the 'First Industrial Nation': The Case of East Anglican Agriculture, 1780–1850." Paper presented at the Conference on Women and Industrialization, Bellagio, Italy, August 8–12, 1983, argues the considerable influence of urban literature on farm women there by 1850.

35. *The Plough, The Loom, and the Anvil* 1 (July 1848):271, 397, 463, for "The Whole Duty of Woman," and p. 65 for "Readings for Mothers and Children."

36. David Lloyd, *Economy of Agriculture: Being a Series of Compendious Essays on Different Branches of Farming* (Germantown, Pa.: P. R. Freas, 1832); Donald B. Marti, "Agricultural Journalism and the Diffusion of Knowledge: The First Half-Century in America," *AH* 54 (1980):34; and Dominic Ricciotti, "Popular Art in Godey's Lady's Book: An Image of the American Woman, 1830–1860," *History of New Hampshire* 27 (1972):22.

37. Sally McMurry, "Progressive Farm Families and Their Homes, 1830–1955," *AH* 58 (1984):330–346, and her "Country Parlors and City Parlors: Conflicting Versions of Nineteenth-Century Women's Culture," Paper delivered at the Berkshire Conference on Women's History, Smith College, 1984.

38. Mary Johnson, "Madame E. I. du Pont and Madame Victorine du Pont Bauduy, the First Mistresses of Eleutherian Mills: Models of Domesticity in the Brandywine Valley during the Antebellum Era," *Working Papers from the Regional Economic History Research Center* 5, nos. 2 & 3 (1982):14–38.

39. Ibid. See also Ruth C. Linton, "The Brandywine Manufacturers' Sunday School: An Adventure in Education in the Early Nineteenth Century," *DH* 20 (1983).

CHAPTER 8

1. Story of ore discovery in Graceanna Lewis, "Notes on her father John Lewis, Jr. & her mother Esther Fussell," in Esther Lewis Papers, RG 5, Series 1, FHLSC.

2. The court case is in Narratives of the Court of Common Pleas, May Term 1824, Registers Court, 5–10–1824, 5–10–1824; Continuance Docket, J.No.8, Feb. T. 1824–No.57, Nov. T. 1824, May Term 1824 and Appearance Docket, No. 20, August 1823–February 1828, *Esther Lewis* v. *Joseph J. Lewis*, CCCH.

3. Family background in Rebecca F. Trimble, "Recollections of Her Father and Mother," and Graceanna Lewis, "Notes on her Father John Lewis & her Mother Esther Fussell," in Esther Lewis Papers, RG 5, Series 1, FHLSC; Agnes Longstreth Taylor, *The Longstreth Family* (Philadelphia: Ferris & Leach, 1909), pp. 111–113. The farm was purchased by John, Sr., from his father-in-law, John Meredith. Grace Meredith and John Lewis married in 1775. John had received cash as his dower, and another son Evan received the Lewis 196-acre farm. Both John and sister Elizabeth received identical sums of money in dower, sixty pounds, indicating strict partible inheritance patterns in this Welsh family. John, Jr., received the entire farm because an older stepbrother Abel was mentally handicapped. Biographical and Genealogical Fan Chart of the Lewis and Fussell families, 1863, and wills in RG 5, Series 1, Esther Lewis Papers, FHLSC. The farm was half of the original three-hundred-acre tract received in 1708. The Lewis family gave each child five hundred dollars or land.

4. Inventory #7471, April 2, 1824, CCCH.

5. Graceanna Lewis, "Notes on her father John Lewis, Jr., and her mother Esther Fussell," RG 5, Series 1, FHLSC. Esther actually cared for two dependents, Abel and Hannah Lewis, but Hannah was provided for out of the John Lewis, Sr., estate and at first was cared for in the Quaker mental asylum. Both survived Esther.

6. Uwchlan Monthly Meeting Women's Minutes, 1818–1829, and Women's Minutes, 1828–1831 (Hicksite), FHLSC; Esther Lewis to Hannah Churchman, 2–10–1838, Esther Lewis Papers, RG 5, Series 2, FHLSC.

7. Children of Ann and Norris are listed in the Uwchlan index, FHLSC.

8. Esther Lewis to Jonathan and Ann Thomas, 8–26–1837, Esther Lewis Papers, RG 5, Series 2, FHLSC.

9. Esther Lewis to Jonathan and Ann Thomas, 8–26–1837, Esther Lewis to S. Fussell, 11–13–1832; Esther Lewis to Rebecca Fussell Trimble, 4–18–1833, Esther Lewis Papers, RG 5, Series 2, FHLSC.

10. Esther Lewis Diary, 4–21–1847, RG 5, Series 4, FHLSC.

11. Esther Lewis to Edwin and Rebecca Fussell, 9–29–1829, Esther Lewis Papers, RG 5, Series 2, FHLSC. See McMurry, "Progressive Farm Families," pp. 330–346 for redesign of farmhouses.

12. The records of ore are in the Esther Lewis Diary, 1830–1833, Esther Lewis Papers, RG 5, Series 4, FHLSC. For Phoenix Iron Works, see Stewart Huston, "The Iron Industry of Chester County," in Nolan, *Southeastern Pennsylvania*, p. 270.

13. Rachel's work described by Esther Lewis to Neal and Elizabeth Hardy, 4–12–1830; Esther Lewis to Jonathan and Ann Thomas, 8–26–37, Esther Lewis to Edwin and Rebecca Fussell, 4–10–1839, 5–29–41, 9–17–42; Graceanna Lewis to Edwin and Rebecca Fussell, 7–23–1841; and Esther Lewis to Graceanna Lewis, 9–12–43 RG 5, Series 2, and Esther Lewis Diary, 1844, RG 5, Series 4, Esther Lewis Papers, FHLSC.

14. Textile processing in Esther Lewis Diary, 9–9–31, 1–17–32, 1–19–32, 1–23–32, 6–16–1842, RG 5, Series 4; and Receipt Book, 1836, RG 5, Series 6, Esther Lewis Papers, FHLSC.

15. Sewing in Esther Lewis to Edwin Fussell, 12–10–1837, RG 5, Series 2 and Esther Lewis Diary, 12–10–1837, 9–9–1838, 11–14–41, RG 5, Series 4, Esther Lewis Papers, FHLSC.

16. Food processing in Esther Lewis Diary, 12–23–31, RG 5, Series 4, Esther Lewis Papers, FHLSC.

17. Purchases scattered through Esther Lewis Diaries for 1830–1839, Esther Lewis Papers, RG 5, Series 4, FHLSC.

18. Inventory #7471, John Lewis, Jr., April 2, 1824, CCCH. Esther Fussell Lewis Book of Appraisement, 1845, in RG 5, Series 4, Esther Lewis Papers, FHLSC.

19. Purchases of books mentioned in Esther Lewis to Rebecca Fussell, 2–1–1837; Esther Lewis to Edwin and Rebecca Fussell, 5–29–41, RG 5, Series 2; Esther Lewis Diary, 5–2–1831, 12–30–1831, 1–8–1835, 7–20–1836, 10–19–1837, RG 5, Series 4, Esther Lewis Papers, FHLSC.

20. Esther Lewis Diary, 3–19–1831, 1–19–1842, RG 5, Series 4; Esther Lewis to Edwin and Rebecca Fussell, 10–4–1838, RG 5, Series 2, Esther Lewis Papers, FHLSC.

21. Daughters' work discussed in Esther Lewis to Solomon Fussell, 6–1–1830, Edwin Fussell, 7–6–1836 and 10–16–1836, and Jonathan and Ann Thomas, 8–26–1837, RG 5, Series 2; Esther Lewis Diary 6–9–1830, 6–23–1830, 7–1–1830, 11–22–1830, 11–27–1830, 2–1–1831, 4–14–1831, 7–11–1831, 8–31–1831, 3–8–1832, 11–16–1832, 6–13–1834, 5–10–1835, 4–19–1836, 11–11–1836, 1–2–1836, 5–29–1836, 7–25–1836, 7–30–1836, RG 5, Series 4, Esther Lewis Papers, FHLSC.

22. References to blacks in Esther Lewis Diary, 1–1824, 2–1824, and strangers 8–12–1836, RG 5, Series 4, Esther Lewis Papers, SFLSC.

23. Graceanna Lewis to Edwin Fussell, 10–16–1836, Esther Lewis to Edwin Fussell,

no date, to Edwin and Rebecca Fussell, 5–29–1841, to Rebecca Fussell, 2–5–1842, RG 5, Series 2, Esther Lewis Papers, FHLSC.

24. Esther Lewis to Edwin and Rebecca Fussell, 9–17–1842, RG 5, Series 2, Esther Lewis Papers, FHLSC.

25. Esther Lewis Diary, 6–26–1831, 7–10–1831, and Esther Lewis Book of Appraisement, 1845, RG 5, Series 4, Esther Lewis Papers, FHLSC.

26. Esther Lewis Diary, 9–30–1832, 5–8–1833, RG 5, Series 4, Esther Lewis Papers, FHLSC.

27. Esther Lewis Diary, 2–4–1831, 2–26–31, 1–8–1832, 1–8–1844, RG 5, Series 4, Esther Lewis Papers, FHLSC. Deborah Jean Warner, *Graceanna Lewis: Scientist and Humanitarian* (Washington, D. C.: Smithsonian Institution Press, 1979).

CHAPTER 9

1. Rosemary Radford Ruether and Rosemary Skinner Keller, *Women and Religion in America: The Nineteenth Century* (San Francisco: Harper & Row, 1981), pp. 208, 211–212.

2. David S. Reynolds, "The Feminization Controversy: Sexual Stereotypes and the Paradoxes of Piety in Nineteenth-Century America," *New England Quarterly* 53 (1980):96–105; Amanda Porterfield, *Feminine Spirituality in America: From Sarah Edwards to Martha Graham* (Philadelphia: Temple University Press, 1980); and Mary P. Ryan, *Cradle of the Middle Class: The Family in Oneida County New York, 1790–1865* (Cambridge: Cambridge University Press, 1981).

3. Judith L. Weidman, ed., *Women Ministers* (San Francisco: Harper & Row, 1981), pp. 1–3; Virginia Lieson Brereton and Christa Ressmeyer Klein, "American Women in Ministry: A History of Protestant Beginning Points," in *Women in American Religion*, ed. Janet Wilson James (Philadelphia: University of Pennsylvania Press, 1980), pp. 171–190.

4. Ruether and Keller, *Women and Religion in America*, p. 208.

5. Leontine T. C. Kelly, "Preaching in the Black Tradition," in *Women Ministers*, p. 72, and Jean M. Humez, " 'My Spirit Eye': Some Functions of Spiritual and Visionary Experience in the Lives of Five Black Women Preachers, 1810–1880," in *Women and the Structure of Society: Selected Research from the Fifth Berkshire Conference on the History of Women*, ed. Barbara J. Harris and JoAnn K. McNamara (Durham, N.C.: Duke University Press, 1984) discuss black women preaching. Mary P. Ryan, "A Women's Awakening: Evangelical Religion and the Families of Utica, New York 1800–1840," *AQ* 30 (1978):602–623; Anne M. Boylan, "Sunday Schools and Changing Evangelical Views of Children in the 1820s," *Church History* 48 (1979):320–333; Anne M. Boylan, "The Role of Conversion in Nineteenth-Century Sunday Schools," *American Studies* 20 (1979):35–47; Anne M. Boylan, "Evangelical Womanhood in the Nineteenth Century: The Role of Women in Sunday Schools," *FS* 4 (1978):62–80; and Elaine J. Lawless, "Shouting for the Lord: The Power of Women's Speech in the Pentecostal Religious Service," *JAF* 96, no. 382 (1983):434–459 all discuss various aspects of evangelical women's preaching. See also Brereton and Klein, "American Women," pp. 174–179; and Porterfield, *Feminine Spirituality*, pp. 51–81.

6. Ann Moore's Journals, FHLSC.

7. Margaret Spufford, *Contrasting Communities: English Villages in the Sixteenth and Seventeenth Centuries* (Cambridge: Cambridge University Press, 1974), pp. 283, 302, 304.

8. "A Lyst of the pasingers abord the Speedwell of London, Robert Lock Master, bound for New England," *New England Historical and Genealogical Register* 1 (1847):132; Elfrida Vipont, *George Fox and the Valiant Sixty* (London: Hamilton, 1975), pp. 128–129, lists twelve of the sixty-six missionaries as women; Frederick B. Tolles, "Atlantic Community of the Early Friends," Supplement, *Journal of the Friends Historical Society* 24 (1952):35–38 lists 25 or 44 percent of the missionaries as women. J. W. Lippincott,

"Society of Friends, Deceased Ministers and Elders, List Compiled from Friends Memoirs and Journals, 1656–1898," FHLSC, and Heiss, *Quaker Biographical Sketches*, p. 62, contain lists.

9. Ibid., p. 62; George Selleck, *Quakers in Boston, 1656–1964: Three Centuries of Friends in Boston and Cambridge* (Cambridge, Mass.: Friends Meeting, 1976), pp. 23, 28; Elizabeth Webb, "Journal of Elizabeth Webb, Called by Her, 'A Short Account of My Viage Into America With Mary Rogers My Companion,' " ed. Frederick B. Tolles and John Beverly Riggs from the original manuscript, 1959, QCHC; Richard Bauman, *Let Your Words Be Few: Symbolism of Speaking and Silence among Seventeenth-Century Quakers* (Cambridge: Cambridge University Press, 1983), pp. 84–94.

10. Margaret Fell, *Womens Speaking: Justified, Proved and Allowed of by the Scriptures, All such as speak by the Spirit and Power of the Lord Jesus* (Amherst, Mass.: New England Yearly Meeting of Friends, 1980, reprint of 1666 London ed.). For other arguments by men who supported women preaching, see Barbara Yoshioka, "Imaginal Worlds: Woman as Witch and Preacher in Seventeenth Century England," (Ph.D. diss., Syracuse University, 1977); Mary Maples Dunn, "Saints and Sisters: Congregational and Quaker Women in the Early Colonial Period," *AQ* 30(1978):595, 597; and Margaret Hope Bacon, *As the Way Opens: The Story of Quaker Women in America* (Richmond, Ind.: Friends United Press, 1980), p. 7.

11. Ibid., p. 13; Rufus M. Jones, *The Quakers in the American Colonies* (New York: Norton, 1966, reprint of 1911 ed.), pp. 113, 231–232; A. L. Morton, *The World of the Ranters: Religious Radicalism in the English Revolution* (London: Lawrence & Wishart, 1970), pp. 70–92, discusses beliefs; Geoffrey F. Nuttall, *Studies in Christian Enthusiasm: Illustrated from Early Quakerism* (Wallingford: Pendle Hill, 1948), pp. 82–85, mentions that Quakers had saved the country from being overrun with ranters. Ida Raming, *The Exclusion of Women from the Priesthood: Divine Law or Sex Discrimination* (Metuchen, N.J.: Scarecrow, 1976), discusses earlier Catholic exclusion.

12. Mary Maples Dunn, "Women of Light," in *Women of America: A History*, ed. Carol Ruth Berkin and Mary Beth Norton (Boston: Houghton Mifflin, 1979), p. 121; Dunn, "Saints and Sisters," p. 598; Heiss, *Quaker Biographical Sketches*, p. 6, mentions a separate meeting at the home of minister Frances Taylor in 1682; see also Bacon, *As the Way Opens*, p. 15.

13. Heiss, *Quaker Biographical Sketches*, pp. 44, 53.

14. Purchase Monthly Meeting, New York, extracts from the minutes and memorandum of Friends traveling in the ministry, 1725–1846, QCHC; Names of Public Friends, Elders and some others visiting monthly meeting, Newport, Rhode Island, 1650–1838, HC; Frederick B. Tolles, "The Transatlantic Quaker Community in the Seventeenth Century," *Huntington Library Quarterly* 14 (1951):246.

15. "Ann Moore's Journal," in *FL*, 14 vols. (Philadelphia: n.p., 1837–1850), 4:289–306; Heiss, *Quaker Biographical Sketches*, p. 122. See also Leslie R. Gray, "Phoebe Roberts' Diary of a Quaker Missionary Journey to Upper Canada," *Ontario History* 42 (1950):7–46, who at fifty-five in 1821 traveled twenty-two hundred miles from Richland, Pennsylvania, to Canada and back, and Janis Calvo, "Quaker Women Ministers in Nineteenth Century America," *QH* 63 (1974):75–93.

16. I have followed Rufus M. Jones, *The Later Periods of Quakerism*, 2 vols. (Wesport, Conn.: Greenwood, 1970, reprint of 1921 ed.), 1:435–487, in this description of the complex split.

17. Lippincott, "Society of Friends, Deceased Ministers and Elders," FHLSC. For London meeting, see J. William Frost, *The Quaker Family in Colonial America: A Portrait of the Society of Friends* (New York: St. Martins, 1973), p. 226.

18. Principal sources for this table are Heiss, *Quaker Biographical Sketches*; John Smith, "Lives of Ministers among Friends," vol. 1, QCHC; Lippincott, "Deceased Ministers," FHLSC; and Journals and Diaries of Quaker women at FHLSC and QCHC.

19. Carol Edkins, "Quest for Community: Spiritual Autobiographies of Eighteenth-Century Quaker and Puritan Women in America," in *Women's Autobiography*, ed. Estelle C. Jelinek (Bloomington: Indiana University Press, 1980), p. 52.

20. Howard H. Brinton, *Quaker Journals, Varieties of Religious Experience among Friends* (Wallingford, Pa., Pendle Hill, 1972), pp. 1–4.

21. Susannah Lightfoot and Elizabeth Daniel in *A Collection of Memorials Concerning Divers Deceased Ministers and Others of the People Called Quakers* (Philadelphia: Crukshank, 1887), pp. 205, 400; "The Life of that faithful servant of Christ, Jane Hoskens," in *FL*, 1:468.

22. Ibid., 1:463.

23. An abstract of the travels with some other remarks of Elizabeth Hudson, QCHC; Ann Roberts and Susanna Morris in Heiss, *Quaker Biographical Sketches*, pp. 176, 186; Mary Emlen in *A Collection of Memorials*, p. 370.

24. Mary Weston diary of religious tour 1750–1751, copy in CCHS.

25. Ann Cooper Whitall Diary, QCHC.

26. "Memoirs of Elizabeth Collins," *FL* 11:472, 452.

27. Ibid., pp. 451–452.

28. "The Life of Jane Hoskens," *FL* 1:469; James Emlen Journal, FHLSC; Herbert A. Wisbey, Jr., *Pioneer Prophetess: Jemima Wilkinson, the Publick Universal Friend* (Ithaca: Cornell University Press, 1964), pp. 4, 14–15. Jemima was expelled from the Friends in 1776 and her father was disowned the following year. Horace Mather Lippincott, *An Account of the People Called Quakers in Germantown, Philadelphia* (Burlington, N.J.: Enterprise, 1923), p. 24, for Priscill Deaves in 1802. For Hannah Barnard, see "Dictionary of Quaker Biography," QCHC; Benjamin Ferris, "An Account of the Separation in the Society of Friends in the Year 1827," RG 5, Ser. 4, Box 12, FHLSC.

29. An abstract of the travels with some other remarks of Elizabeth Hudson, QCHC.

30. Diary of Esther (Roberts) Hunt Collins, QCHC. She was an elder and sister-in-law of minister Elizabeth Collins.

31. "Memoirs of Elizabeth Collins," *FL* 11:452; for Ruth Walmsley, see Ibid., 6:77; Journal of Ann Shipley, QCHC.

32. For Rachel Barnard, see *Memorials Concerning Deceased Friends* (Philadelphia: Kite, 1842), p. 3; "An Acount of Part of the Travels of Susanna Morris," HSP; "An Abstract of the Travels with Some Other Remarks of Elizabeth Hudson," QCHC; Mary Weston Diary of religious tour, 1750–1751, CCHS.

33. "The Life of Jane Hoskens," *FL* 1:470, 472; "An Account of Part of the Travels of Susanna Morris," HSP; Elizabeth Wilkinson, "Journal of a Religious Visit to Friends in America, 1761–63," QCHC.

34. Jane Hoskins to Elizabeth Hudson, 6–1–1749, Elizabeth Hudson to Jane Hoskins, 6–8–1749, to "Dear Companion," 5–28–1749, in "An abstract of the Travels with Some Other Remarks of Elizabeth Hudson," QCHC.

35. Tolles and Riggs, "A Short Account," QCHC; "An Abstract of the Travels with Some Other Remarks of Elizabeth Hudson," QCHC and Ann Moore's Journals, FHLSC, describe harassment in detail.

36. Elizabeth Wilkinson, Journal of a Religious Visit to Friends in America, 1761–1763, QCHC; Sarah Cresson Diary, QCHC; *Biographical Sketches and Anecdotes of Members of the Religious Society of Friends* (Philadelphia: Tract Association of Friends, 1870), p. 355, for Elizabeth Foulke. Richard Bauman, "Speaking in the Light: The Role of the Quaker Minister," in *Explorations in the Ethnography of Speaking*, ed. Richard Bauman and Joel Sherzer (London: Cambridge University Press, 1974), pp. 144–160.

37. Diary of Esther (Roberts) Hunt Collins, QCHC; Memoir of Ruth Richardson, Richardson Family Papers, WML; "An Abstract of the Travels . . . Elizabeth Hudson,"

QCHC; *Memorials Concerning Deceased Friends* (Philadelphia: Solomon W. Conrad, 1821), p. 420; *Biographical Sketches and Anecdotes of Members of the Religious Society of Friends*, p. 120.

38. Dana Greene, *Lucretia Mott: Her Complete Speeches and Sermons* (New York and Toronto: Mellen, 1980).

39. "An Abstract of the Travels... Elizabeth Hudson," QCHC.

40. "The Life of Jane Hoskens," *FL* 1:471; Elizabeth Wyatt in Heiss, *Quaker Biographical Sketches*, p. 170; "Memoirs of Elizabeth Collins," *FL* 11:464; Diary of Sarah Cressan, QCHC.

41. "An Account... Susanna Morris," HSP.

42. Elizabeth Wilkinson Journal, QCHC; Ann Cooper Whitall Diary, QCHC.

43. Ibid.

44. "Verses on Elizabeth Ashbridge by MW," Copy books of Hannah Minshall, 1795–1796, Series 8, Box 25, Painter Collection, FHLSC.

45. Center Minutes for Women's and Men's Meetings, FHLSC.

46. Rachel Coope, "Journal 1805," FHLSC; Ann Mifflin, "Religious Visit to the Indians of Upper Canada in 1802, and in 1803 to the Seneca on the Allegheny River," HSP; "Some Account of a Visett divers friends made to the Indians at the time of the Treaty at Easton taken by one of the Company as follows—1761," HSP, which included Susannah Hatton, Susannah Brown, and others; Hannah Yarnall, "Visits of Friends in Canada, 1803," Richardson Papers, FHLSC.

47. Elizabeth Webb, "A Short Memorial of the Dealings of God with Me in the Days of My Youth," QCHC; Diary of Ester (Roberts) Hunt Collins, refers to 1787 meeting for black people, QCHC; *Two Hundred Fifty Years of Quakerism at Birmingham, 1690–1940* (West Chester, Pa.: Birmingham Friends, 1940), pp. 21–23; Ann Wilson to Davis Wilson, 6–20–1821, reported meeting with "people of colour" in Flushing, New York, Corbett-Wilson Papers, WML; Journal of Ann Shipley, reported blacks attending yearly meeting in Richmond, Virginia, in 1821, QCHC.

48. Ann Shipley, for example, refused when asked to "discourse," Journal of Ann Shipley, QCHC; lime quarries visit in Center Minutes for Women's and Men's Meetings, 8–1–1825, FHLSC; Margaret Judge meeting with sailors and their families and to New York penitentiary discussed in Ann Wilson to David Wilson, 6–20–1821, Corbett-Wilson Papers, WML.

49. Gerda Lerner, *The Grimké Sisters from South Carolina: Pioneers for Women's Rights and Abolition* (New York: Schocken, 1971), p. 118.

CHAPTER 10

1. Linda K. Kerber, *Women of the Republic: Intellect and Ideology in Revolutionary America* (Chapel Hill: University of North Carolina Press, 1980), p. 199; Nancy F. Cott, *The Bonds of Womanhood: "Woman's Sphere" in New England, 1780–1835* (New Haven and London: Yale University Press, 1977), p. 121; and Norton, *Liberty's Daughters*, p. 204, discuss the ideology of republican mothers. Kathryn Kish Sklar, *Catharine Beecher: A Study in Domesticity* (New York: Norton, 1973), p. 97.

2. Anne Firor Scott, "The Ever Widening Circle: The Diffusion of Feminist Values from the Troy Female Seminary, 1822–1872," *History of Education Quarterly* 19 (Spring 1979):3, 7; Saul Mulhern, *A History of Secondary Education in Pennsylvania*, (Harrisburg: the author, 1933), p. 378.

3. Alan Tully, "Literacy Levels and Educational Development in Rural Pennsylvania, 1729–1775," *PH* 39 (1972):301–312. Linda Auwers, "Reading the Marks of the Past: Exploring Female Literacy in Colonial Windsor, Connecticut," *Historical Methods* 13 (1980):204–214, argues for a higher literacy in some areas of Massachusetts by the mid-

eighteenth century than does Kenneth Lockridge in *Literacy in Colonial New England* (New York: Norton, 1974).

4. Sydney V. James, "Quaker Meetings and Education in the Eighteenth Century," *QH* 51 (1962):97, discusses problems of implementing educational goals.

5. George S. Brookes, *Friend Anthony Benezet* (Philadelphia: University of Pennsylvania Press, 1937), p. 53.

6. Diary of Sarah Cresson, 1789–1892, QCHC; "Ruth Walmsley," in *FL* 6:75–82.

7. Quoted in Monica Kiefer, "Early American Childhood in the Middle Atlantic," *PMHB* 68 (1944):17.

8. Thomas Woody, *Early Quaker Education in Pennsylvania* (New York: Columbia University, 1920), p. 181. Five women are mentioned in a 1784 list of teachers.

9. Ibid., p. 211; School for Black People, Overseers' Minutes, FHLSC. Brissot de Warville, *New Travels*, p. 217, referred to girls learning writing as well as reading, religion, spinning, and needlework. Debra L. Newman, "Black Women in the Era of the American Revolution in Pennsylvania," *JNH* 61 (1976):288, lists girls learning writing in 1800 at the abolitionists' school.

10. Joseph Hawley, Diary, 1796 and School Account, 1796, CCHS.

11. Providence Meeting Subscription List, Painter Collection, Series 5, Box 3, FHLSC.

12. Joan N. Burstyn, *Victorian Education and the Ideal of Womanhood* (Stowa, N.J.: Barnes & Noble, 1980), p. 11, discusses this tension for women in education. I have chosen to emphasize the liberating aspects for middle-class women because the documents most clearly reveal this attitude. Thomas Woody, *A History of Women's Education in the United States*, 2 vols. (New York and Lancaster, Pa.: Science Press, 1929), 1:202, for Philadelphia Free Institute; *Mirror of the Times*, 4–2–1816, 2:2, describes the Harmony Society; and Edwin W. Rice, *The Sunday-School Movement 1780–1917 and the American Sunday-School Union, 1817–1917* (Philadelphia: American Sunday-School Union, 1917), the literacy movement. Among Quakers there is little of the rhetoric of piety and submissiveness described by Kieth Melder, "Woman's High Calling: The Teaching Profession in America, 1830–1860," *American Studies* 13 (Fall 1972):19–32.

13. Esther Fussell to John Lewis, Jr., 4–20–1817, Esther Lewis Papers, FHLSC. Wells, "Quaker Marriage Patterns," pp. 415–442.

14. Mary Johnson, "Madame Rivardi's Seminary in the Gothic Mansion," *PMHB* 105 (1980):3–39; "Victorine du Pont: Heiress to the Educational Dream of Pierre Samuel du Pont de Nemours," *DH* 19 (1981):88–105; and "Antoinette Brevost: A Schoolmistress in Early Pittsburgh," *WP* 15 (1980):151–168. Cott, *Bonds of Womanhood*, p. 121, talks of the "vicissitudes of fortune" being recognized as a reason for education in tracts after 1819.

15. Sarah Dewess, *History of Westtown Boarding School 1799–1899* (Philadelphia: Sherman, 1899), pp. 90, 99.

16. Ibid., pp. 20, 45.

17. Ibid., pp. 54, 69, 70, 92, and Helen G. Hole, *Westtown through the Years, 1799–1942* (Westtown, Pa.: Westtown Alumni Association, 1942), p. 55.

18. Rebecca Budd, Diary, and Olive Baily to Rest Swayne, 11–9–1808, CCHS. John C. Appel, "Significance of the Quakers in Public Education in Pennsylvania, 1801–1860" (Master's Thesis, University of Maryland, 1943), pp. 86, 89, 91, 93, for teacher training.

19. Rachel Painter to Sarah Painter, 10–17–1821, Painter Collection and Martha Sharpless, Journal of Philadelphia Meeting, 4–1804, FHLSC.

20. *Catalog of Westtown through the Years* (Westtown, Pa.: Westtown Alumni Association, 1945), pp. 340, 341, 351.

21. Kimberton School, Clipping File, CCHS.

22. Esther Lewis to Edwin and Rebecca Fussell, 4–10–1839, Esther Lewis Papers, FHLSC. Graceanna Lewis, "Kimberton Boarding School for Girls: Abby Kimber," *Friends'*

Intelligencer 52 (1895):615–616, 633–634, 662–663, and Alice Lewis, "Abby Kimber," *Friends' Intelligencer* 53 (1896):734.

23. Student memoirs and letters are in the Howard M. Jenkins Papers, RG 5, S. C. Haines Manuscripts, and the Palmer Papers, RG 5, especially series of letters from Pattie to Sallie T. Palmer in 1843 from Kimberton, FHLSC.

24. The joy of learning is strongly conveyed by an unidentified writer at Wilmington Boarding School to Esther Lewis, 12–8–18, Esther Lewis Papers, FHLSC. Information on *Sharon Female Seminary* from undated brochure Sharon Female Seminary, Painter Collection, Box 6, Folder 16, and Rosa D. Weston, "Quaker Sharon," Misc. Mss. c.1902, FHLSC; Eleanor Wolf Thompson, *Education of Ladies, 1830–1860* (New York: King's Crown, 1947), pp. 64–65; Mary S. Bartram, "John Jackson," *Friends' Intelligencer* 53 (1896):155–156; and Deborah Jean Warner, "Science Education for Women in Antebellum America," *Isis* 69 (1978):59.

25. Rachel Painter to Elizabeth Painter, 11–30–1822, Painter Collection, FHLSC.

26. Rachel Painter to Minshall Painter, 6–20–1815 and 1–3–1817, Letter Book, Index A. Letters to Minshall and Enos Painter, 1820–1863, Painter Papers, PLTA.

27. Rachel Painter to Minshall Painter, 1–16–1820, PLTA.

28. Rachel Painter to Minshall Painter, 3–15–1823, PLTA.

29. Superintendent of Common Schools of Pennsylvania, *Eighth Annual Report* (Harrisburg, 1842). Over 25 percent of 1,041 names in a list of payments to teachers for teaching paupers from 1805 to 1834 were female. Since most of the payments were from the 1830s and the earliest names were usually female, women had already moved into teaching in Chester County in large numbers before the public school system was started. The 286 names of women teachers are listed in the appendix to George T. Hanning, "A Historical Survey of the Schools in Chester County Prior to 1834" (Master's Thesis, Temple University, 1937). Shur, "Emergence of the Free, Common Schools," p. 191, notes that the first woman superintendent in the state in 1865 was from West Chester and that in the 1872 election, all the women elected to serve as directors of schools came from Quaker districts.

30. There are no comparable statistics for New Castle County, Delaware, but scattered indications of women's lower employment and wage rates are in Federal Writers' Project, University of Delaware, Special Collections, Newark, Delaware, vol. 2, especially pp. 278–285. Employment statistics for Delaware County are in Minshall Painter, "Statistics of the Populations of Middletown Township and Delaware County," PLTA. District schools employed six times as many males as females and paid them one-third more in 1850. Opposition to schooling of Afro-Americans included using the courts to invalidate an endowment for a black college in Wilmington. The Delaware *Gazette and Watchman* noted the decision with approval on 12–12–1834, 2:1. Indenture records, DHR, reflect the growing opposition to literacy. Harold C. Livesay, "Delaware Negroes, 1865–1915," *DH* 13 (1968):89, erroneously states that all blacks were illiterate. This was not true for blacks in New Castle County.

31. Harold B. Hancock, "Mary Ann Shadd: Negro Editor, Educator, and Lawyer," *DH* 15 (April 1973):187–194; Elsie M. Lewis, "Mary Ann Shadd Cary," in James, *Notable American Women,* 1:300–301.

32. Biographical information in Gulielma F. Alsop, "Ann Preston," ibid., 3:96–97 and Eliza E. Judson, *Address in Memory of Ann Preston, M.D.* (n.p., 1873), FHLSC.

33. Ann Preston to Hannah Darlington during 1837 to 1854 describes her activities. Quotes from letters of 1–9–1851 and 2–9–1850, CCHS.

CHAPTER 11

1. Hannah M. Darlington to Jacob Painter, February 6, 1852, Painter Collection, Series 1, Box 6, Folder 29, FHLSC.

2. U.S., Manuscript Census, 1850, lists both Darlingtons and Painters.

3. Elizabeth Cady Stanton, Susan B. Anthony, and Matilda Joslyn Gage, *History of Woman Suffrage*, 3 vols. (Rochester, n.p., 1889), 1:821, has a list of Pennsylvania women signing the Worcester call.

4. Margaret Hope Bacon, *Valiant Friend: The Life of Lucretia Mott* (New York: Walker, 1980), p. 144; Bayard Taylor quoted in Marie Hansey-Taylor, ed., *Life and Letters of Bayard Taylor*, 2 vols. (Boston: Houghton Mifflin, 1885), 1:4.

5. Ellen DuBois, "Women's Rights and Abolition: The Nature of the Connection," in *Antislavery Reconsidered: New Perspectives on the Abolitionists*, ed. Lewis Perry and Michael Fellman (Baton Rouge and London: Louisiana State University, 1979), pp. 239–251, is a recent restatement of the connection. For cresting of petition activity, see Judith Wellman, "Women and Radical Reform in Antebellum Upstate New York: A Profile of Grassroots Female Abolitionists," in *Clio Was a Woman: Studies in the History of American Women*, ed. Mabel E. Deutrich and Virginia C. Purdy (Washington: Howard University Press, 1980), pp. 113–127, and Gerda Lerner, *The Majority Finds Its Past: Placing Women in History* (New York: Oxford University Press, 1979), pp. 112–201.

6. Peggy A. Rabkin, *Fathers to Daughters: The Legal Foundations of Female Emancipation* (Westport, Conn.: Greenwood, 1980), pp. 1–13.

7. Norma Basch, *In the Eyes of the Law: Women, Marriage and Property in Nineteenth-Century New York* (Ithaca: Cornell University Press, 1982), pp. 119–156.

8. W. J. Rorabaugh, *The Alcoholic Republic: An American Tradition* (New York: Oxford University Press, 1979), pp. 36–43, discusses Quaker opposition from 1706; Elizabeth Levis, "Some friendly advice & Cautions / Wherein are Some things of weighty Concern / Recommended to the Serious Consideration of the professors of the Holy Truth, 1761," FHLSC; Heiss, *Quaker Biographical Sketches*, p. 318. See Marietta, *Reformation of American Quakerism*, pp. 107–108, for movement.

9. David Ferris quoted in J. William Frost, *The Quaker Origins of Antislavery* (Norwood, Pa.: Norwood, 1980), p. 185; Tully, "Patterns of Slaveholding," pp. 284–305; Alice Jackson Lewis described in Futhey and Cope, *History of Chester County*, p. 630.

10. Thomas E. Drake, *Quakers and Slavery in America* (New Haven: Yale University Press, 1950), pp. 42–77; Merle Gerald Brouwer, "The Negro as a Slave and as a Free Black in Colonial Pennsylvania" (Ph.D. diss., Wayne State University, 1973), pp. 212, 282; Arthur Zilversmit, *The First Emancipation: The Abolition of Slavery in the North* (Chicago and London: University of Chicago Press, 1967), p. 53.

11. For black churches, see Baldwin, "Invisible Strands," pp. 66–108, and Oblinger, "Freedom Foundations: Black Communities in Southeastern Pennsylvania Towns," pp. 7, 9–10.

12. Ibid., pp. 121–122; Charles Blockson, "A Black Underground Resistance to Slavery, 1833–1860," *Pennsylvania Heritage* 4 (1971):29–33.

13. Larry Gara, *The Liberty Line: The Legend of the Underground Railroad* (Lexington: University of Kentucky Press, 1961), and "Friends and the Underground Railroad," *QH* 51 (1962):3–19.

14. William Carpenter Patten, "The Anti-Slavery Movement in Chester County" (Master's Thesis, University of Delaware, 1963), pp. 24–32.

15. James A. McGowan, *Station Master on the Underground Railroad: The Life and Letters of Thomas Garrett* (Moylan, Pa.: Whimsie, 1977), p. 121; Lewis account is in her diaries of January and February 1834, Esther Lewis Papers, RG 5, Series 4, FHLSC.

16. Escape of Harris in undated fragment of a letter signed Mother, Esther Lewis Papers, RG 5, Series 2, FHLSC.

17. Graceanna Lewis, "Recollections of Underground Activities at the Lewis Farm," Esther Lewis Papers, RG 5, Series 1, FHLSC.

18. W. V. Hensel, *The Christiana Riot and the Treason Trials of 1851* (Lancaster, Pa.: New Era, 1911).

19. Ira V. Brown, "Cradle of Feminism: The Philadelphia Female Anti-Slavery Society, 1833–1840," *PMHB* 32 (1965):153–165, and Patten, "Anti-Slavery Movement," pp. 39–41.

20. Graceanna Lewis to Rebecca and Edwin Fussell, 8–31–1838, Esther Lewis to Rebecca Fussell, 2–1–1837, Esther Lewis Papers, RG 5, Series 2, FHLSC. Chandler's significance is evaluated by Blanche Glassman Hersh, *The Slavery of Sex: Feminist-Abolitionists in America* (Urbana: University of Illinois Press, 1978), pp. 7–10.

21. Elizabeth Chandler, "The Kneeling Slave," "Letters on Slavery: To the Ladies of Baltimore," are in Benjamin Lundy, *The Poetical Works of Elizabeth Margaret Chandler: With a Memoir of Her Life and Character* (Philadelphia: Lemuel Howell, 1836), pp. 50, 43–48.

22. Rebecca Lewis to Mother and Sisters, 12–12–1837, Esther Lewis Papers, RG 5, Series 2, FHLSC.

23. Ibid.

24. Esther Lewis to Edwin and Rebecca, 5–29–1841, Esther Lewis Papers, RG 5, Series 2, FHLSC.

25. Philadelphia petition of 1836 is in Records of the U.S. House of Representatives, Library of Congress Collection, NA, RG 233, Box 79.

26. Ibid., Box 110.

27. Petitions from Pennsylvania are in NA, RG 233, Box 79, 90, 94, 109, 110. Women had petitioned earlier on private matters, but this was the first time they used the right for public political purposes. Even the right to become naturalized was exercised by very few women in the early nineteenth century. The Regional National Archives, Philadelphia, RG 21, contains declarations of intent and petitions of naturalization from only a handful of women before 1834.

28. Ibid.

29. Wellman, "Women and Radical Reform," pp. 113–127; Lerner, *The Majority Finds Its Past*, pp. 112–201.

30. *Friends Intelligencer and Journal* 53 (2–22–1896, 2–15–1896):108–110, 125–126, based on a letter from Benjamin Fussell to Dr. Edwin Fussell, 5–23–1838, five days after the burning; PA, RG 7, Box 22, Folder 22, 46, 74.

31. Hewett, *Women's Activism*, p. 99, and Brown, "Cradle of Feminism," p. 161.

32. NA, RG 233, Box 150, 151.

33. Minutes of the Lundy Antislavery Society, CCHS.

34. Patten, "Anti-Slavery Movement," pp. 100–102.

35. PA, RG 7, Box 22, Folder 62 contains ten petitions from the city and county of Philadelphia. The background of the petitions is discussed in Charles W. O'Brien, "The Growth in Pennsylvania of the Property Rights of Married Women," *American Law Register* 49 (September 1901):524–530; Charles W. Dahlinger, "The Dawn of the Woman's Movement: An Account of the Origin and History of the Pennsylvania Married Woman's Property Law of 1848," *Western Pennsylvania Historical Magazine* 1 (1918):73–78; Ira V. Brown, "The Woman's Rights Movement in Pennsylvania, 1848–1875," *PH* 32 (1965):153–165.

36. Brinton, "Friends of the Brandywine Valley," p. 85; Tavern Licenses, CCCH; and Marietta, *The Reformation*, p. 107.

37. Union Temperance Society, Ms. 12887, CCHS.

38. Hewett, *Women's Activism and Social Change*, pp. 99, 112–113; Esther Lewis to Rebecca Fussell, 2–5–1842 and to Graceanna Lewis, 5–28–1842, Esther Lewis Papers, RG 5, Series 2, FHLSC. For additional information on temperance, see Asa Earl Martin, "The Temperance Movement in Pennsylvania Prior to the Civil War," *PMHB* 49 (1925):195–230.

39. Jed Dannenbaum, "The Origins of Temperance Activism and Militancy among

American Women," *JSH* 15 (1981):235–252, and Ian R. Tyrrell, "Women and Temperance in Antebellum America, 1830–1860." *Civil War History* 28 (1982):128–152.

40. PA, RG 7, Box 22, Folders 43–44, 73.

41. Ibid., Folder 73.

42. Hannah M. Darlington to Elizabeth Cady Stanton, 2–6–1881, in Stanton, Anthony, and Gage, *History of Woman Suffrage*, 1:344.

43. Address at December 1848 meeting in ibid, 1:345.

44. Rorabaugh, *Alcoholic Republic*, pp. 142, 146, 169, 178, 187, judges 1790–1830 as peak of consumption, the 1820s as a period of increase of binges, and the 1830s as a period of decline in consumption. He argues the movement preceded immigration of the 1840s, but the earlier immigration was also accompanied by social dislocation and alcoholism. See Henry H. Bisbee and Rebecca Bisbee Colesar, *Martha, 1808–1815: The Complete Furnace Diary and Journal* (Burlington, N.J.: Bisbee, 1976), for an impressive account of drunkenness and physical abuse of women. Rural battering may have been common. On the 1854 vote, see Martin, "Temperance Movement," p. 226.

45. Esther Lewis diary 9–28–1815, 12–30–1830, Esther Lewis Papers, RG 5, Series 2, FHLSC.

46. Rebecca Lewis to Mother and Sisters, 12–12–1837, Dr. Edwin Fussell to Solomon Fussell, 3–20–1838, Graceanna Lewis to Rebecca and Edwin Fussell, 8–31–1838, Esther Lewis Papers, RG 5, Series 2, FHLSC.

47. Stanton, Anthony, and Gage, *History of Woman Suffrage*, 1:821; Woman's Rights Meeting in Upper Providence, 9–13–1851, to appoint delegates is PLTA.

48. Hannah M. Darlington to Jacob Painter, February 6, 1852, Painter Collection, Series 1, Box 6, Folder 29, FHLSC.

49. Hannah M. Darlington to Jacob Painter, February 22, 1852, FHLSC.

50. *American Republican and Chester County Democrat*, April 20, 1852, CCHS.

51. *Jeffersonian*, May 25, 1852, June 2, 1852.

52. *American Republican*, June 8, 1852.

53. Stanton, Anthony, and Gage, *History of Woman Suffrage*, 1:833–834.

54. Ibid., 1:342. Olivia Harris, "Households and Their Boundaries," *History Workshop* 13 (Spring 1982):151, reminded me of the political importance of the sibling tie in egalitarian political movements.

55. Stanton, Anthony, and Gage, *History of Woman Suffrage*, 1:833–834.

56. Albert John Wahl, "The Congregational or Progressive Friends in the Pre-Civil-War Reform Movement," (Ph.D. diss., Temple University, 1951), pp. 47–62. Biographical information comes from U.S., Manuscript Census, 1850, and various scattered biographical references in CCHS.

57. Ibid.

58. In her Rochester study, Hewett found women's rights activists, whom she calls "ultras," had smaller families than either benevolent or evangelical women in 1842–1848. The family size of over a third of her small sample was unknown, however. She did not look at the gender of children living at home in 1850, *Women's Activism*, p. 156, and Tables 6, 14, and 21.

59. Ann Preston, December 19, 1852, excerpt, 9–10–1874, clipping file, CCCHS.

EPILOGUE

1. Female Benevolent Society Records; Female African School Society Records, HSD. W. Emerson Wilson, *Diaries of Phoebe George Bradford, 1832–1839* (Wilmington: Historical Society of Delaware, 1976), pp. 19, 35.

2. *West Chester Directory* (West Chester: n.p., 1857), *Wilmington Directory* (Wilmington, Del.: n.p., 1845) and *Wilmington Directory* (Wilmington, Del.: n.p., 1853); Dunn

and Bradstreet Records, Delaware, Mary Dixon, microfilm copy at EMHL: M.C.L. Reminiscences of Wilmington, Delaware, 1815–1828, University of Delaware, Special Collections.

3. Priscilla Durham listed as recipient of clothes in Female Benevolent Society Records, HSD; Susan Harley in U.S., Manuscript Census, 1850.

4. Board of Health Records, DHR; U.S., Manuscript Census, 1850, listed "ladies of the town" as well as women textile factory workers. Strike is in *Blue Hen's Chickens*, 12–31–1852.

INDEX